Walter Michaeli

Plastics Processing

An Introduction

Hanser Publishers, Munich Vienna New York
Hanser/Gardner Publications, Inc., Cincinnati

The Author:
Professor Dr. Walter Michaeli, IKV, Pontstraße 49, 52062 Aachen, Germany

Original title:
Einführung in die Kunststoffverarbeitung by W. Michaeli

Translated by Dr. Kurt Alex, 20 Derman Street, Rumford, RI 02916-1906, USA

Distributed in the USA and in Canada by
Hanser/Gardner Publications, Inc.
6600 Clough Pike, Cincinnati, Ohio 45244–4090, USA
Fax: (513) 527–8950
Phone: (513) 527-8977 or 1–800–950–8977

Distributed in all other countries by
Carl Hanser Verlag
Postfach 86 04 20, 81631 München, Germany
Fax: +49 (89) 98 12 64

Library of Congress Cataloging-in-Publication Data
Michaeli, Walter.
[Einführung in die Kunststoffverarbeitung. English]
Plastics processing : an introduction / Walter Michaeli.
p. cm.
Includes bibliographical references and index.
ISBN 1-56990-144-9
1. Plastics. I. Title
TP1120.M48713. 1995.
668.4–dc20 95-3748

Die Deutsche Bibliothek – CIP-Einheitsaufnahme
Michaeli, Walter:
Plastics processing : an introduction / Walter Michaeli. -
Munich ; Vienna ; New York ; Hanser ; Cincinnati :
Hanser/Gardner, 1995
Einheitssacht.: Einführung in die Kunststoffverarbeitung <engl.>
ISBN 3-446-17572-5

Typeset in Great Britain by Techset Composition Ltd., Salisbury
Printed and bound in Germany by Schoder Druck GmbH & Co. KG, Gersthofen

Preface

To give his students their first comprehensive look at the fascinating world of plastics and its processing techniques, *Professor Georg Menges*, the former head of the Institute of Plastics Processing in Aachen, Germany, created lecture notes on "Plastics Processing". His text proved its value time and again. To further develop and to make this text available for the broad industrial plastics community, we were asked to create a book that would present the full spectrum of plastics processing in a form suitable for self-study and also easily understood by individuals without a scientific education.

The result is a book that provides an introduction and thorough survey of the major plastics processing methods, how they operate and their underlying principles.

This book represents a joint effort. Special thanks are extended first to *Professor Menges* for his initial work and the fact that we are permitted to build on it. Further thanks are extended to the staffs of the two German plastics institutes, SKZ and IKV, for their contributions to the success of this book. The plastics associations supported us generously—technically and financially—in converting the idea of this book into reality. *Dr. Kurt Alex* has translated this text into English using all his experience and knowledge of plastics technology. We also thank Carl Hanser Verlag, who took our text and made it into a book, which I hope the reader will appreciate and enjoy.

Professor Dr. W. Michaeli

Contents

1 Introduction

Human use of organic, high-molecular-weight materials such as wood, textiles, or leather has been documented since early times. This use was restricted to the processing of existing materials, however. The conscious transformation of natural materials into those classified as plastics today did not begin until a century ago. It has only been since the 1930s, however, that plastics have achieved major economic importance, after Herrmann Staudinger developed in the early 1920s the model for the structure of plastics and the resulting possibilities for synthesis.

Worldwide growth of the plastics industry did not begin until after World War II. Initially, coal served as the basis for the starting materials until the less expensive petroleum became available in the mid-1950s. The advantage of this conversion was that refinery fractions which had been worthless to date and which were produced during the cracking of crude oil could now be utilized meaningfully. Thus began the rapid growth in plastics production, parallel to the growing use of gasoline and heating oil, and it did not slow down until the oil crisis of 1973. While the growth rate in the use of plastics has slowed since then, these materials continue to show above-average, dynamic development.

Today, plastics applications are found in almost all areas of life. Polymer materials are employed with great success for both mass-produced products used in the fields of construction, packaging, agriculture, household appliances, and leisure goods as well as for higher valued articles such as electronics, automotive components, precision instruments, and even high-tech applications in the aircraft and aerospace industries.

In opening new fields of application, the substitution of classical (metallic) materials with plastics is becoming more and more pronounced. The myriad ways in which the material properties can be influenced have permitted plastics to advance to the status of a "tailor-made material." A further advantage is that plastics are a typical "mass-production material." The reasons for this can be found in the easy processibility into parts with even complicated shapes in very few operations—with injection molding, for instance, only one operation—and this at a relatively low temperature level, which means low energy costs. In spite of the ability to be easily machined manually, piece part production is found in plastics processing to only a limited extent, or machining in a manner similar to metals starts with semi-finished products.

It is not without reason that plastics have experienced such rapid development. Plastics are materials with an extremely wide spectrum of possibilities. Their properties can be summarized briefly as follows:

- Plastics are light. Their density varies between 0.8 and 2.2 g/cm^3. They are thus lighter than metals and ceramics.
- Plastics exhibit a wide, variable spectrum of mechanical properties. They can be soft and elastic as well as hard and rigid.
- Plastics can be processed simply and economically at low temperatures into complex parts that often require no secondary finishing. Plastics production and processing require relatively little energy input. The thermal stability of plastics, however, is limited. Plastics

are good thermal and electrical insulators, although in a few applications, the exact opposite is required and can also be achieved within limits.

- Plastics are often transparent and can be colored as desired.
- Plastics have high chemical resistance.
- Plastics are permeable (i.e., permit permeation and diffusion). This is not always desired, but it differs with the resin. Thus, there are applications where this exact property is necessary, e.g., in membranes for desalination of sea water.
- Plastics can be reused and recycled by means of a number of different methods.

The use of polymeric materials is successful only when the particular features of the materials are taken into consideration. Especially when substituting materials, a change from the strategies used to date, e.g., with regard to material selection and component part design, to a "plastics-oriented" approach is essential. This requires a thorough study of this materials group, including resin synthesis and the synthetic methods employed, material properties, and material behavior as well as the variety of primary processing and secondary finishing methods. It is the objective of this book to provide a simple and easy-to-understand introduction to all of these.

Bibliography for Chapter 1

Menges, G.: Werkstoffkunde Kunststoffe, 3rd ed., Carl Hanser Verlag, München, Wien, 1990

2 Structure and Classification of Plastics

2.1 Structure of Plastics

The first plastics were developed in the second half of the nineteenth century through chemical transformation of high-molecular-weight natural materials (Galalith, celluloid, rayon), partly as substitutes for the expensive and highly valued natural products. In 1905 Baeckeland succeeded in producing resin from formaldehyde and phenols in such a manner that the resulting products were materials that could be processed. The first attempts at producing a synthetic elastomer ("rubber," methyl rubber)—once again as a substitute for the natural rubber no longer available because of the blockade during World War I—were already conducted in Germany during World War I. The development of artificial fibers can be traced above all to the work of Carothers (USA), who succeeded in producing polyamides that could be spun into fibers from dicarboxylic acids and diamines.

True progress, however, was hindered by the lack of any basic knowledge of the structure of these materials. For a long time, it was generally assumed that even materials such as cellulose, rubber, or vinyl polymers consisted basically of molecules with relatively low molecular weights. The special properties of these materials were considered a consequence of the agglomeration of molecules into colloid particles, so-called *micelles*.

It was only after the energetic pioneering work of the German chemist Herrmann Staudinger, who introduced the term macromolecule in 1922, that actual progress began. Staudinger recognized that high polymers consist of molecules with very high molecular weights, the so-called *macromolecules*, in which the individual carbon atoms are bonded to one another in exactly the same manner as in any other low-molecular-weight organic substance. For this work, which laid the foundation for polymer science, Staudinger received the Nobel Prize in 1953.

2.1.1 Polymer Synthesis

The production of plastics is based on three reaction processes:

1. polymerization,
2. polycondensation,
3. polyaddition.

2.1.1.1 Polymerization

Polymerization describes a *chain reaction* in which unsaturated molecules are linked to form macromolecules (polymers). No by-products are formed in the course of the reaction.

The low-molecular-weight starting molecules are called *monomers*. They have double or triple bonds that can be broken. The formation of polyethylene is presented as the simplest example:

$n \cdot CH_2{=}CH_2 \rightarrow$	$\lbrack CH_2 \rbrack_n$ ———	Degree of polymerization (number of monomers forming the macromolecule)
Monomer	Structural unit of the polymer	

During polymerization, each monomer is converted into a structural unit of the polymer chain. The number of structural units n that form a molecule chain is called the *degree of polymerization*. The resulting molecules form the polymer, i.e., the plastics resin.

The expression "chain reaction" has nothing to do with the fact that in this case molecule chains are formed, but rather it describes the kinetics of the reaction, which progresses in three successive phases or reactions steps:

- the starting (initiating) reaction,
- the chain propagation reaction,
- the terminating (interrupting) reaction.

Depending on the type of reactive particle that triggers the polymerization, a distinction is made between the following types of polymerization:

- free-radical polymerization,
- cationic polymerization,
- anionic polymerization,
- metal-complex polymerization.

(The reaction steps for metal-complex polymerization cannot be formulated unambiguously, since it has not been possible to date to fully clarify the reaction sequence for this type of polymerization.)

Free-Radical Polymerization

To start the polymerization, so-called initiators decompose upon input of energy into very reactive free radicals, which then react with the monomers, e.g., peroxides such as hydrogen peroxide (H_2O_2) or dibenzoyl peroxide. The reaction sequence is as follows (Fig. 2.1.1):

1. Starting reaction, i.e., the initiators decompose into their free radicals upon input of energy.
2. Propagation reaction:
   a) the radicals now react with double-bonded carbon,
   b) the new group reacts further with other double bonds.
3. Interruption, i.e., polymerization comes to a stop because of:
   a) reaction of two radical ends on polymer chains,
   b) reaction of a radical end on a polymer chain with the radical of an initiator,
   c) transfer of a hydrogen atom from one macroradical to another,
   d) elimination of a bonding hydrogen atom.

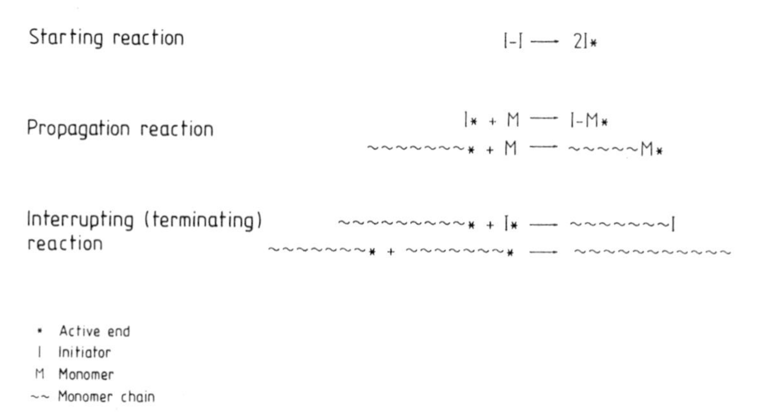

Figure 2.1.1 Polymerization.

Ionic Polymerization

In contrast to free radical polymerization, where the growing chain has a reactive location (radical) at one end, positively charged ions (cations) or negatively charged ions (anions) are always present at the end of the chain during ionic polymerization.

Cationic Polymerization

Catalysts for cationic polymerization include acids (e.g., sulfuric acid, H_2SO_4, and phosphoric acid, H_3PO_4) and the so-called Friedel–Crafts catalysts (e.g., boron trifluoride, BF_3, and aluminum chloride, $AlCl_3$) in combination with water. These substances release protons, which attract electrons as a consequence of their positive charge. The cationic polymerization of isobutylene with boron trifluoride/water to form polyisobutylene is as follows (Fig. 2.1.2):

1. Starting reaction, i.e., ion formation via an acidic catalyst.
2. Propagation reaction.
3. Chain interruption; this can also occur as a result of the reaction with an anion (e.g., from the catalyst) if arbitrary termination of the reaction is not induced through the addition of bases or similar substances.

Anionic Polymerization

Catalysts for anionic polymerization include strong bases (e.g., sodium hydroxide, NaOH, and potassium hydroxide, KOH). In practice, strong bases based on organometallic compounds (e.g., sodium alcoholates, R-ONa, and sodium amide, $NaNH_2$) are often employed. These substances are *nucleophilic*. They satisfy with their free electron pair the positively polarized C atom on the monomers. This produces an anion that generates a growing chain with other monomers. Chain termination always occurs through the addition

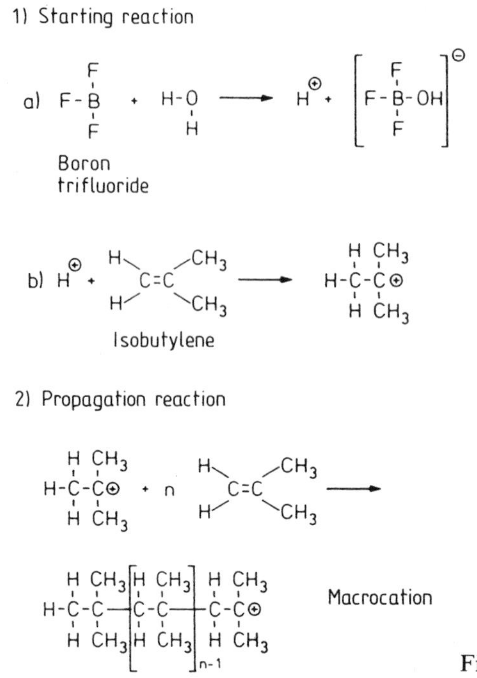

1) Starting reaction

a) $\mathrm{BF_3} + \mathrm{H_2O} \longrightarrow \mathrm{H}^{\oplus} + [\mathrm{F{-}BF_2{-}OH}]^{\ominus}$

Boron trifluoride

b) $\mathrm{H}^{\oplus} + \mathrm{H_2C{=}C(CH_3)_2} \longrightarrow \mathrm{H_3C{-}C^{\oplus}(CH_3)_2}$

Isobutylene

2) Propagation reaction

$\mathrm{H_3C{-}C^{\oplus}(CH_3)_2} + n\,\mathrm{H_2C{=}C(CH_3)_2} \longrightarrow$

$\mathrm{H_3C{-}C(CH_3)_2}{-}\left[\mathrm{CH_2{-}C(CH_3)_2}\right]_{n-1}{-}\mathrm{CH_2{-}C^{\oplus}(CH_3)_2}$ Macrocation

Figure 2.1.2 Cationic polymerization.

of a proton, e.g., from a solvent molecule. Anionic polymerization of acrylic acid methyl ester with sodium amide forms polymethacrylate (Fig. 2.1.3).

Metal Complex Polymerization

The great importance of this polymerization method is that it produces very regularly structured chains. For instance, in contrast to the polyethylene produced via free radical polymerization (low-density polyethylene), the polyethylene produced by this method (high-density polyethylene) has none or only a few chain branches and thus a higher density as the result of a greater degree of crystallization (Fig. 2.1.4). So-called *metal complex compounds*, e.g., a mixture of titanium trichloride and triethyl aluminum, serve as *catalysts*.

The mechanism of this type of polymerization is not yet known in detail.

It must be assumed on the basis of the various polymerization possibilities as well as on the variety of chain termination reactions, which proceed simultaneously, that the plastics resin thus synthesized does not consist only of chains with a uniform length (molecular weight). Plastics always exhibit a more or less broad molecular weight distribution. This has a significant effect on the property profile of the material.

The Technology of Polymerization

In order to start chain formation, a certain activation energy, usually introduced in the form of heat or radiation, is necessary. The following technical polymerization methods are distinguished (see also Table 2.1).

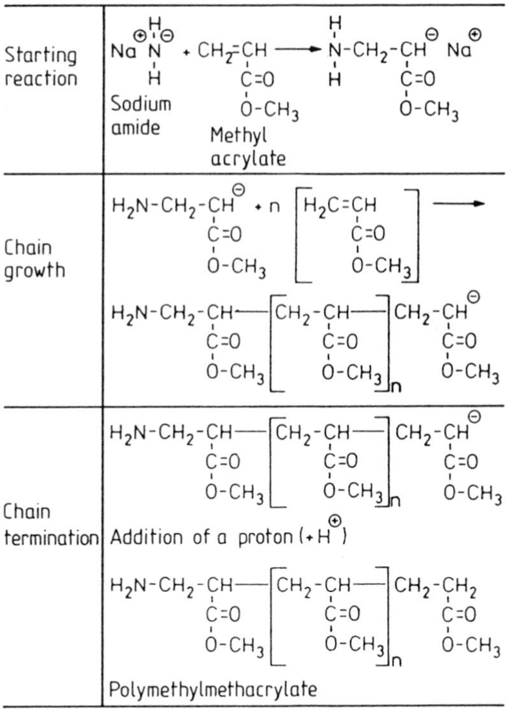

Starting reaction	$Na^{\oplus}\,{}^{\ominus}NH_2$ (Sodium amide) + $CH_2{=}CH{-}C({=}O){-}O{-}CH_3$ (Methyl acrylate) $\longrightarrow$ $H_2N{-}CH_2{-}CH^{\ominus}({-}C({=}O){-}O{-}CH_3)\ Na^{\oplus}$
Chain growth	$H_2N{-}CH_2{-}CH^{\ominus}({-}C({=}O){-}O{-}CH_3) + n\,[H_2C{=}CH({-}C({=}O){-}O{-}CH_3)] \longrightarrow$ $H_2N{-}CH_2{-}CH({-}C({=}O){-}O{-}CH_3){-}[CH_2{-}CH({-}C({=}O){-}O{-}CH_3)]_n{-}CH_2{-}CH^{\ominus}({-}C({=}O){-}O{-}CH_3)$
Chain termination	$H_2N{-}CH_2{-}CH({-}C({=}O){-}O{-}CH_3){-}[CH_2{-}CH({-}C({=}O){-}O{-}CH_3)]_n{-}CH_2{-}CH^{\ominus}({-}C({=}O){-}O{-}CH_3)$ Addition of a proton ($+H^{\oplus}$) $H_2N{-}CH_2{-}CH({-}C({=}O){-}O{-}CH_3){-}[CH_2{-}CH({-}C({=}O){-}O{-}CH_3)]_n{-}CH_2{-}CH_2({-}C({=}O){-}O{-}CH_3)$ Polymethylmethacrylate

Figure 2.1.3 Anionic polymerization.

HDPE

Linear molecules
Approximately 4 to 10 short side chains per 1000 C atoms

LDPE

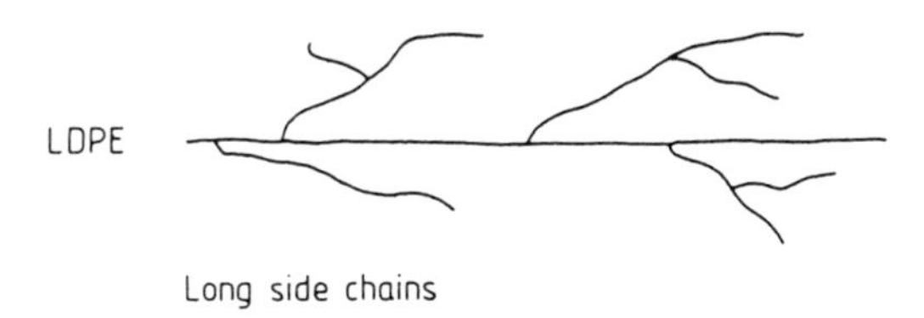

Long side chains

LLDPE

Linear molecules
Approximately 10 to 35 short side chains per 1000 C atoms

Figure 2.1.4 Molecular structure of various PE types.

Table 2.1 Overview of Technically Important Polymerization Methods

Polymerization method	Features
Block (mass) polymerization	Direct polymerization of monomers under pressure and temperature (without additional fluid)
Solution polymerization	The monomers are dissolved in a liquid. The polymer is polymerized. The polymer remains in a solution
Precipitation polymerization	The monomers are dissolved in a liquid. The polymer is insoluble and precipitates
Emulsion polymerization	The monomers are finally dispersed in a so-called emulsifier and polymerized. The polymer remains dispersed
Bead (suspension) polymerization	The monomers form small beads upon being stirred in water and polymerize. The polymer precipitates

a) *Block Polymerization (Mass Polymerization)*

Here the monomers, together with initiators and catalysts, react spontaneously with one another, without the addition of liquids, under certain reaction conditions, such as pressure and temperature. Controlling the heat of reaction is, naturally, the greatest problem. Heat removal is controlled by limiting the size (cross section) of the tubular reaction chambers. The most important reaction of this type is the high-pressure polymerization of polyethylene, where gaseous ethylene is polymerized under pressures of approximately 2000 bar (29,000 psi) and temperatures of 200°C in high-pressure reactors—pipes with extremely good temperature control. Styrene, methacrylic methyl ester (methyl methacrylate), butadiene, and vinyl chloride also polymerize in this manner.

In a few cases, mass polymerization can also take place directly in molds (*in situ*). However, the difficulties associated with temperature control and the shrinkage (which often creates difficulties even for other areas of plastics processing) pose problems that have limited large-scale use of this method to only especially high-grade polymerizates, for instance, organic glasses of polymethyl methacrylate (PMMA).

b) *Solution Polymerization*

Here the monomers as well as the resulting polymers are dissolved in a liquid for the purpose of simpler temperature control. Polyvinyl acetate (PVA), high-density polyethylene, and polyacrylic acid, for instance, are produced by this means. After polymerization, the solvent must be removed, possibly by means of evaporation. Often the resulting polymer solutions are employed directly, e.g., for adhesives.

c) *Precipitation Polymerization*

In this method, the monomers are dissolved and polymerized in a solvent. In contrast to the monomer, the resulting polymer is insoluble and precipitates. This process is more economical than solution polymerization, since it is faster and the solvent can be used again immediately. High-density polyethylene and polyisobutylene, for instance, are produced by this method.

d) *Emulsion Polymerization*

In emulsion polymerization the monomers are emulsified in water. Fatty acid soaps and the like serve as emulsifiers. Monomer droplets are enclosed by soap micelles. The water-soluble radicals diffuse into these. Polymerization takes place in the droplets. The polymerizate itself appears as an aqueous dispersion. A subsequent separation process yields a very finely powdered material (e.g., the so-called E-PVC). The subsequent separation process is not always necessary. In these cases, the dispersion is utilized directly as a coating, adhesive, or textile finish. Emulsifier residues in the polymerizate, however, can sometimes have a negative effect.

e) *Suspension Polymerization (Bead Polymerization)*

In suspension polymerization, the monomers and initiators are finely dispersed in water by appropriately strong mechanical stirring. The problem of removing the emulsifier does not exist in this case, because the polymerizate occurs in the form of small beads, which can be easily separated from the water. High-grade polymerizates are produced. Under certain circumstances, the beads can serve directly as pellets. Polyacrylonitrile and polyvinyl acetate, for instance, are produced in this manner.

2.1.1.2 Polycondensation

Polycondensation describes a *multiple-step reaction* during which macromolecules (polycondensates) are formed while by-products (e.g., water, ammonia, alcohol) are released. The starting materials do not need to contain any double bonds as in polymerization, but rather they must have so-called *functional groups*, i.e., groups of atoms that are especially reactive; these include, for instance:

- hydroxyl group $-OH$
- carbonyl group $>C=O$
- carboxyl group $-C{\overset{O}{\underset{OH}{\lessgtr}}}$
- amino group $-N{\overset{H}{\underset{H}{<}}}$

Linear, branched, and cross-linked polycondensates are produced, depending on whether the monomers are bi-, tri-, or higher order compounds, i.e., they have two, three, or more functional groups (Fig. 2.1.5):

Polycondensation represents a true chemical equilibrium reaction. If the by-products are not removed, the reaction stops when equilibrium is established between the monomers, on the one hand, and the polymers and by-products, on the other. Thus, the reaction can be controlled in stages, i.e., it can be interrupted and restarted. This is often utilized.

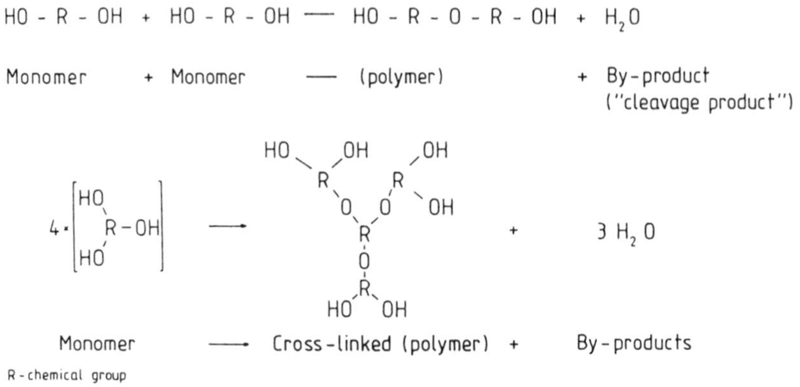

HO - R - OH + HO - R - OH — HO - R - O - R - OH + H_2O

Monomer + Monomer — (polymer) + By-product ("cleavage product")

4 · [(HO)(HO)R-OH] ⟶ (HO)(OH)R-O, (OH)(OH)R-O, R-O-R(HO)(OH) + 3 H_2O

Monomer ⟶ Cross-linked (polymer) + By-products

R - chemical group

Figure 2.1.5 Polycondensation.

The removal of the by-products, however, often creates technical problems, e.g., the formation of bubbles that may make the use of polycondensation products unsuitable, e.g., for casting or laminating. The most important groups of polycondensates are:

- melamine resins,
- urea resins,
- phenolic resins,
- polyamides,
- polycarbonates,
- polyesters,
- silicones.

2.1.1.3 *Polyaddition*

As is the case for polycondensation, starting materials for polyadditions must also contain particularly *reactive functional groups.* In contrast to condensation, however, no by-products are formed, but rather only high-molecular-weight *polyadducts* are created through the "addition" of the monomers (Fig. 2.1.6).

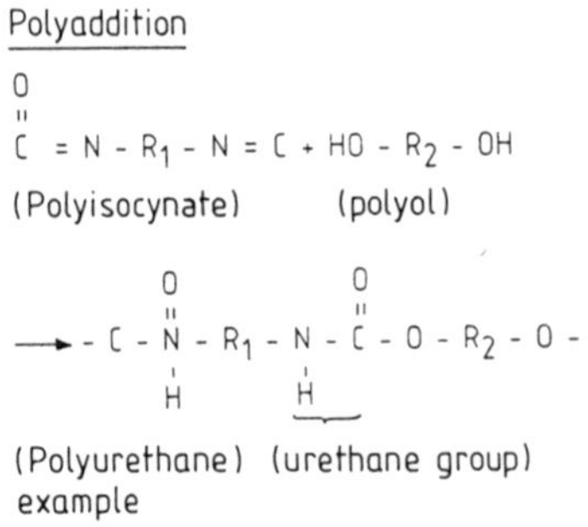

Polyaddition

O=C = N - R_1 - N = C=O + HO - R_2 - OH

(Polyisocynate) (polyol)

⟶ - C(=O) - N(H) - R_1 - N(H) - C(=O) - O - R_2 - O -

(Polyurethane) example (urethane group)

Figure 2.1.6 Synthesis of polyurethane as an example of polyaddition.

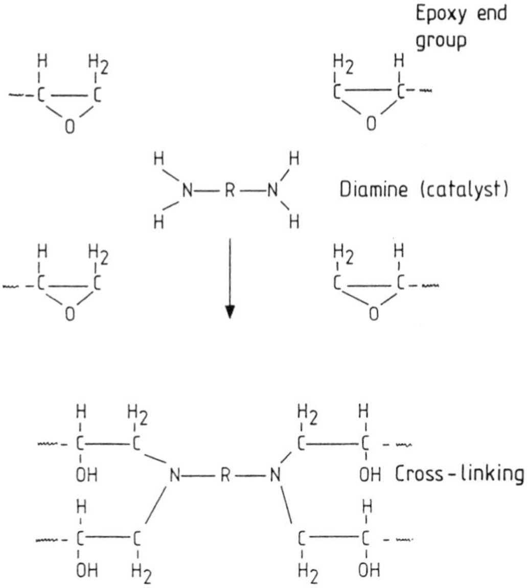

Epoxy end group

Diamine (catalyst)

Cross-linking

Figure 2.1.7 Cross-linking reaction of epoxy resins.

Characteristic of this type of reaction is that hydrogen atoms migrate from the functional group on one of the starting molecules to that on one of the others. The cause is usually polarization of the functional group, creating a positive charge on the carbon atom. Typical polyadducts include:

- polyoxymethylenes,
- polyurethanes,
- epoxy resins.

Polycondensation and polyaddition reactions can only occur between different types of molecules, which must furthermore have at least two functional, i.e., reactive, regions in their molecular structure. In polycondensation and polyaddition, linking of the monomers occurs through the bonding of reactive end groups, while in polymerization a double bond is broken to form two single bonds.

If one of the described reactions involves only bifunctional monomers, i.e., the monomers can form only two bonds, only linear chain molecules can be produced. If the monomers are multifunctional, i.e., can form several bonds, the chains also cross-link to one another, and a network results (Fig. 2.1.7).

2.1.2 Bonding Forces in Polymers

Within a plastics resin, *primary* and *secondary valence forces* act among the macromolecules and determine the properties. The primary valence forces involve the chemical *bonds between atoms*, which are also called *covalent* or *electron pair* bonds. Depending on the number of bonds between two atoms, *single*, *double*, and *triple bonds* are distinguished.

In contrast to the primary valence forces, the secondary valence forces, which are also called *intermolecular forces* or *van der Waal's forces*, are only physical in nature. Intermolecular forces are those forces between macromolecules that result from the distribution and motion of the electrons. Depending on their origin, they can be assigned to the following groups:

a) dispersion forces,

b) dipole forces,

c) hydrogen bonds.

a) Dispersion Forces

Dispersion forces are the attractive forces that generally act in materials. The bonding energy of these dispersion forces is inversely proportional to the sixth power of the distance between molecules and lies in the range of <10 kJ/mole. As might be expected, the dispersion forces are especially large in crystalline regions, because here the molecules have the closest possible packing, i.e., the shortest distance to one another. This is the origin of the noticeable strength of plastics, above all in the highly stretched condition.

b) Dipole Forces

If two atoms with different electronegativities form a covalent bond, a *dipole* results. A dipole is characterized by a permanently asymmetric electron distribution, i.e., the more electronegative atom attracts the electrons more strongly and as a consequence is negatively polarized; the other atom in the bond is electron-deficient and thus positively polarized.

Dipole forces are weakened by thermal motion, because the dipoles can be turned from their original positions. When side chains become entangled, dipoles are often brought into closer proximity, and their action extends over greater intermolecular distances.

Polar molecules are formed when the concentrations of charge are displaced in molecular chains. When the macromolecules have a symmetrical structure (e.g., polyethylene), the concentrations of charge lie along the axis of the molecule; as a result, they are nonpolar.

c) Hydrogen Bonds

Hydrogen bonds exist between dipoles consisting of a positively polarized hydrogen atom and a more electronegative atom, e.g., O, Cl, N, and F. Because of the pronounced action of the dipole forces, the bonding enthalpy of hydrogen bonds is especially high compared to that of other secondary valence forces. The resulting attractive forces determine the strength of many polymers, such as cellulose, proteins, polyamides, and polyurethanes (Fig. 2.1.8).

The strength of the secondary valence forces, and thus their importance in regard to molecular properties, is strongly dependent on the size, structure, and arrangement of the macromolecules as well as on the type of molecule and other external factors. Upon heating, for instance, the secondary valence forces are progressively overcome. This occurs at a characteristic temperature for each molecule.

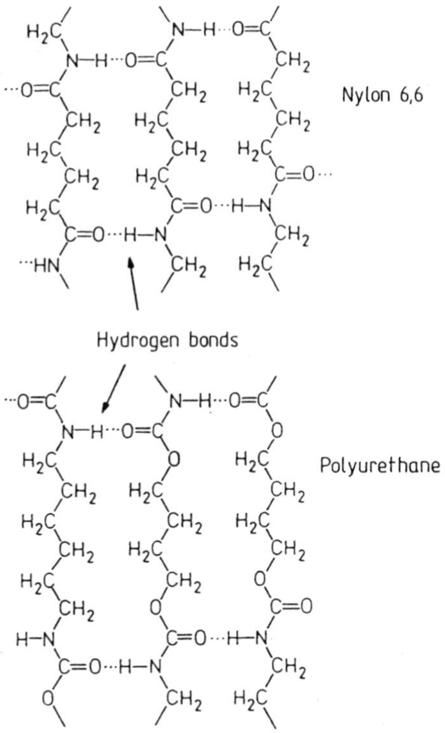

Nylon 6,6

Hydrogen bonds

Polyurethane

Figure 2.1.8 Hydrogen bonds in polyamide (nylon) 6,6 and polyurethane.

2.2 Classification of Plastics

Plastics are classified into certain material groups on the basis of the structure and the bonding mechanism between the macromolecules comprising the plastic. Depending on the type of macromolecule (Fig. 2.2.1), the following classifications are used:

- linear chain molecules — thermoplastics
- branched chain molecules — thermoplastics
- weakly cross-linked chain molecules — elastomers
- highly cross-linked chain molecules — thermosets

2.2.1 Thermoplastics

Plastics whose macromolecules have *linear* or *branched chains* are called *thermoplastics*. The individual molecular chains are held together exclusively by *secondary bonding forces*. The magnitude of the secondary forces depends on the type and number of branches or side chains, among other things. If the molecules have only a few branches, i.e., short and few side chains, the individual molecules can lie close together. The process of forming the dense packing arrangement of molecules is also called *crystallization*. Because of the length of the

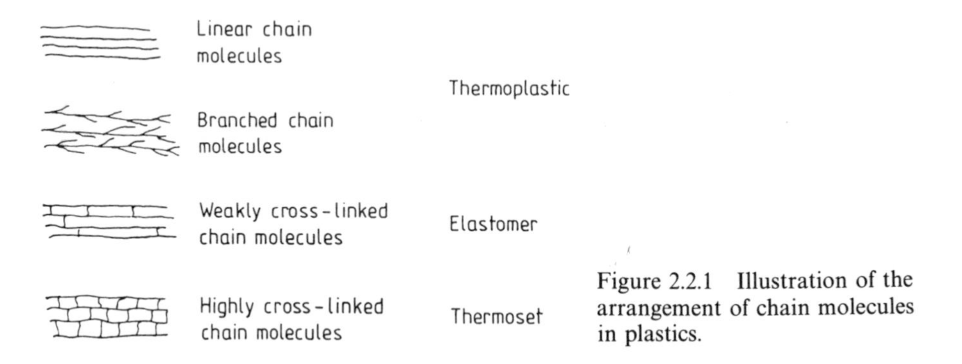

Figure 2.2.1 Illustration of the arrangement of chain molecules in plastics.

molecule chains, however, which can become entangled in and around one another during synthesis, complete crystallization does not occur in plastics. For this reason, one speaks of *semi-crystalline plastics*. The semi-crystalline thermoplastics are never crystal clear in the unpigmented state, but rather are always somewhat translucent or milky because of the scattering of light at the boundaries of the crystallites (crystalline regions).

In contrast, plastics whose monomeric structural units exhibit complex structures (e.g., the benzene ring on the C atom in polystyrene), or whose molecular chains are highly branched and whose side chains are long, can, because of their irregular structure, never achieve the close packing found in crystallites. In this case, the molecule chains are coiled around one another and entangled as in a cotton wad. They solidify in an *amorphous* manner as do the inorganic glasses. For this reason, these plastics are also classified as *amorphous thermoplastics*. Since they are always crystal clear in the unpigmented condition, these thermoplastics are also called *synthetic* or *organic glasses*. Fig. 2.2.2 shows a schematic representation of the arrangement of the macromolecules for a semi-crystalline and for an amorphous plastic.

Under normal use conditions (room temperature), thermoplastics are brittle (amorphous thermoplastics) or tough and elastic (semi-crystalline thermoplastics). With increasing temperature the secondary bonding forces between the molecule chains are weakened as a result of the increased thermal vibration. The individual molecule chains can be displaced with respect to one another so that the plastic exhibits elastic material behavior. If the temperature is raised further, the individual molecule chains slide past one another, and the

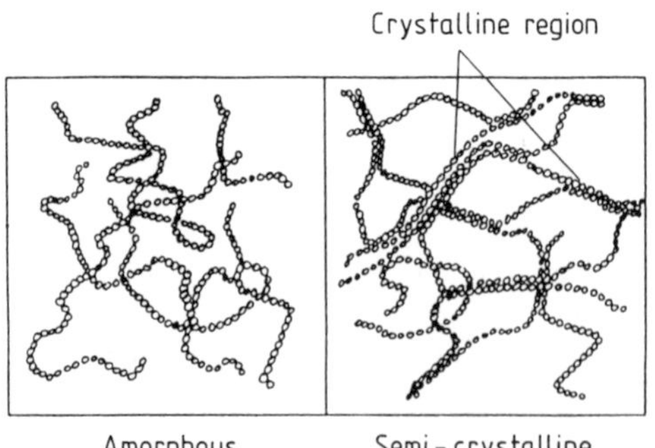

Figure 2.2.2 Structure of an amorphous semi-crystalline.

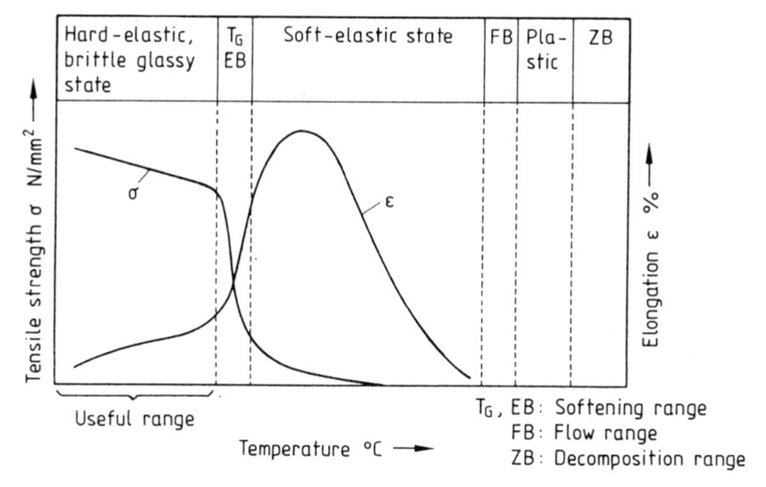

Figure 2.2.3 Tensile strength and elongation of an amorphous thermoplastic.

material is transformed to the state of plastic flow. Transformation from hard and elastic to the plastic state can be repeated indefinitely by heating and cooling for thermoplastics as long as the temperature at which the macromolecules chemically decompose is not exceeded. Because of the mobility of their molecular chains, thermoplastic resins are fusible, weldable, soluble, and swell.

Observing the deformation behavior of an amorphous thermoplastic (Fig. 2.2.3), it is found that the material is hard and brittle at room temperature. With increasing temperature, the strength of the material decreases and, at the same time, the elongation increases. Upon exceeding the *glass transition temperature*, at which the previously "frozen" macromolecules become mobile, the strength drops rapidly, while the elongation increases abruptly. In this temperature range, the plastic is in a rubbery or highly elastic condition. With a further increase in temperature, the plastic exceeds the *flow temperature* and enters the melt state.

The flow temperature is defined as that temperature at which the viscosity is low enough to permit processing on commercially available machinery in order to permit primary conversion. The higher the molecular weight of the polymer, for instance, the higher is the flow temperature because of the greater degree of molecular entanglement.

If the temperature is increased further, the material thermally degrades, i.e., even primary valence bonds are broken. The material then loses its mechanical properties very rapidly.

With the aid of the deformation behavior, the difference between a semi-crystalline and amorphous thermoplastic can be illustrated very clearly. In Fig. 2.2.4 the behavior of a semi-crystalline thermoplastic is shown as a function of the temperature. As already mentioned above, the molecules of a semi-crystalline thermoplastic assume a crystalline structure in certain regions, while the intermediate regions exhibit amorphous structures. Two phases—amorphous and crystalline—thus exist in the plastic next to one another. Below the glass transition temperature T_G, the amorphous regions of the material are solidified so that the plastic is hard and very brittle. Within this temperature range, the plastic is useless for practical applications. Upon exceeding the glass transition temperature,

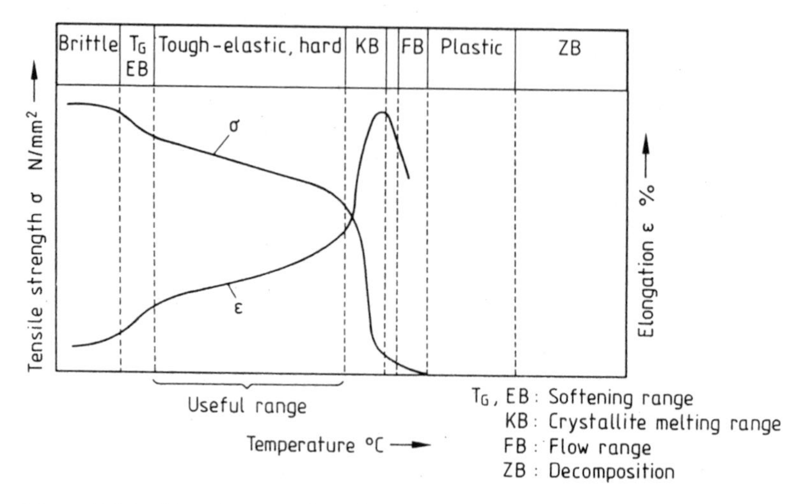

Figure 2.2.4 Tensile strength and elongation of a semi-crystalline thermoplastic.

the mobility of the molecular chains in the amorphous regions increases first. Because of the higher secondary valence forces, the molecular chains in the crystalline phase cannot yet move. The glass transition temperature of common semi-crystalline thermoplastics often lies below room temperature. For practical applications, the plastic is thus in a condition where the amorphous regions are softened, while the crystalline phase is still rigid. Accordingly, the plastic simultaneously exhibits toughness (as a result of the partial mobility) and strength (as a result of the crystallites).

With increasing temperature, the mobility of the molecular chains in the amorphous phase continues to increase. Upon exceeding the crystallite melting temperature, the bonding forces within the crystalline regions become too weak to prevent displacement or sliding of the molecular chains. The crystalline regions begin to melt.

If the molecular weight is high, a single amorphous phase now forms. The long molecules, which were ordered in the crystallites below the crystallite melting temperature, exhibit greater mobility, i.e., a highly viscous melt results. With a further temperature increase, molecular motion becomes so great and the viscosity so low that the flow temperature is exceeded (see Fig. 2.2.4). Thus, a region with rubberlike elasticity exists between the crystallite melting and flow temperatures for high-molecular-weight polymers.

If, however, the polymer has a low molecular weight, the plastic exhibits low viscosity and easy flow upon exceeding the crystallite melting temperature. The region of rubberlike elasticity disappears, since the flow temperature of the amorphous phase is already exceeded. This is the case with most commercially available thermoplastics.

It can be said in summary that thermoplastics are based on long molecular chains that can be linear or branched. The individual molecular chains are held together by secondary bonding forces. For this reason, a thermoplastic can always be remelted. In addition, a thermoplastic swells, can be welded, and is soluble. Thermoplastics with disordered, entwined chains are called amorphous. They exhibit brittle behavior in the solid state and are crystal clear (in the unpigmented condition). Thermoplastics with ordered structures

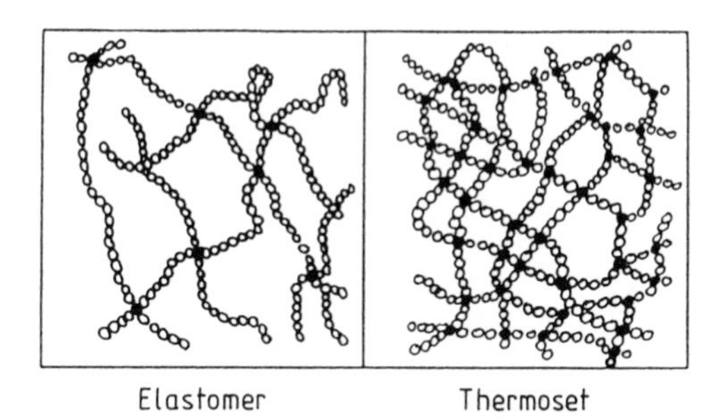

Figure 2.2.5 Structure of cross-linked plastics: elastomer and thermoset.

(crystalline phase) and disordered (amorphous) regions next to one another are called semi-crystalline. They are usually tough and elastic in the region between the glass transition and crystallite melting temperatures and always somewhat translucent when unpigmented.

2.2.2 Thermosets and Elastomers

Besides the group of thermoplastics that are based on linear and/or branched chain molecules, there are groups of plastics in which the individual molecules are bonded to one another by cross-links. These bonds are also called *cross-linking sites* and the materials are classified as *cross-linked plastics.*

Fig. 2.2.5 shows two different molecular structures, both of which are characteristic for cross- linked plastics. In one case, the molecular chains are irregularly distributed and have only relatively few cross-links to neighboring molecules. Plastics with such a low degree of cross-linking are classified as *elastomers*, or materials with rubberlike elasticity, since at room temperature they exhibit for the most part fully reversible deformation behavior. Because of the cross-linking, the individual molecules have only limited mobility with respect to one another. Elastomers are thus neither fusible nor soluble. To a certain degree, however, elastomers can swell.

In the second case, the molecular chains also exhibit an irregular arrangement. Compared to the elastomer structure, however, this structure has considerably more cross-links between the individual molecules. Plastics based on such highly cross-linked molecular chains are called *thermosets.* The macromolecules, which are linked to one another via chemical valence bonds, are also called *network polymers.* These highly cross-linked plastics are very hard and brittle at room temperature and, in comparison to thermoplastics, exhibit considerably less softening. As is also the case for elastomers, they are neither fusible nor soluble. In contrast to the elastomers, however, they exhibit only slight swelling because of the *high degree of cross-linking.*

The various physical states of cross-linked thermoplastics can be explained easily with the aid of the torsional vibration test. In Fig. 2.2.6, the shear modulus, a measure of the stiffness of a plastic, is plotted as a function of temperature for various cross-linked plastics. In the temperature range below the *glass transition temperature*, the plastic is hard and brittle regardless of its degree of cross-linking.

The shear modulus of weakly cross-linked polymers drops also upon exceeding the glass transition temperature, so that the plastic exhibits only limited stiffness. In contrast to a plastic with no cross-linking, however, the weakly cross-linked plastic retains this stiffness

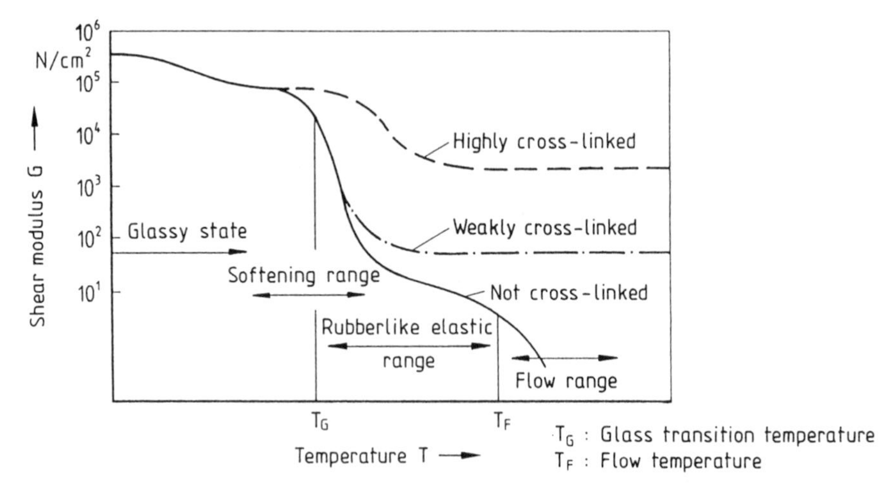

Figure 2.2.6 Shear modulus curves for cross-linked and non-cross-linked polymers.

even as the temperature increases further. The reason for this behavior is the cross-linking sites in the polymer, which prevent sliding of the individual molecules past one another. Thus, the plastic will not melt, but rather will decompose with a further temperature increase (*decomposition temperature* T_Z).

If the plastic is highly cross-linked, its stiffness decreases only slightly, even in the softening range. As a result of the many cross-linking sites between the individual molecules (network polymer), the mobility of the macromolecules with respect to one another is very severely limited. As are the weakly cross-linked polymers, the highly cross-linked plastics are also infusible. Upon reaching the decomposition temperature, this plastic also degrades.

It can be said in summary that cross-linked plastics are divided into two groups. The weakly cross- linked plastics are classified as elastomers. They are neither fusible nor soluble, but they do swell. Under use conditions, or at room temperature, they exhibit rubberlike elasticity. The highly cross-linked plastics are called thermosets. They are neither fusible nor soluble and hardly swell. At room temperature, they are hard and brittle. Their mechanical properties change little at elevated temperatures. For this reason, thermosets are also classified as temperature-resistant plastics.

2.2.3 Copolymers and Polymer Blends

There are also plastics that are based on not only one monomer, but on various monomers. Depending on the manner in which the various monomers are combined, these plastics are called *copolymers* or *polymer blends*. The objective in this case is to combine and utilize the positive properties of each monomer or polymer. For instance, by incorporating elastomeric components, the impact strength of a plastic can be increased.

The term copolymer is used to described plastics whose molecular chains contain several different monomers. In most cases, two different monomeric structural units are involved

Type 1
Random macromolecule

Type 2
Alternating macromolecule

Type 3
Macromolecule composed of blocks

Type 4
Homogeneous chain with grafted side chains

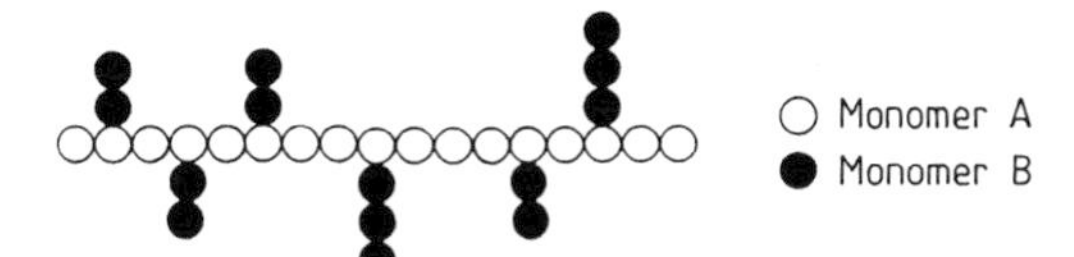

Figure 2.2.7 Types of copolymers.

(*bipolymers*); however, three or four different monomers (*ter-* or *quarter-polymers*) can also be incorporated into the molecular chain. With the copolymers, there are various possibilities for arranging the individual monomers with respect to one another (Fig. 2.2.7). Different macromolecules are produced, depending on the type of monomers and course of the polymerization process. In a random macromolecule, the individual monomeric structural units are distributed randomly along the molecular chain. In an alternating copolymer, a structural unit comprising one monomer alternates regularly with another monomer. In the third case, a block based on one monomer is attached to a block of the second monomer end to end. These are called *block copolymers*. If the two monomers are added to the reaction process in succession, the resulting plastics are called *graft copolymers*. In this case, the macromolecule consists of a chain of the one monomer onto which shorter chains of the other monomer are attached, i.e., grafted.

The other group of plastics, which is formed by mixing different polymers, is called a *polymer blend*. Naturally, the properties of the polymer blend are strongly dependent on the quality and type of mixing of the polymers. If both polymers are compatible with one another, *homogeneous* blends can be produced by mixing the two components in the melt state. As an example, the mixing of two polyamide (nylon) grades (PA 6 with PA 10) is mentioned here. Blends of partially compatible polymers, such as polyethylene and polypropylene, for instance, are called *single-phase mixtures*. *Multiple-phase mixtures* represent the case most commonly found in actual practice. These are so-called *heterogeneous* blends of incompatible polymers. A well-known example of such a multi-phase blend is PP/EPDM (polypropylene/ethylene polypropylene terpolymer). Here the rubber appears in the form of small spheres embedded in a polypropylene matrix.

Additive	Function
Soluble additives	Chemically acting processing and application aids (stabilizers and the like) Physically acting processing aids (lubricants and the like) Plasticizing additives
Insoluble solids	Coloring additives (pigments and the like) Reinforcing additives (tensile strength) Stiffening and hardening additives (modulus of elasticiity) Impact-modifying additives
Insoluble gases	Volume-increasing blowing agents

Figure 2.2.8 Classification of fillers and additives for plastics.

2.2.4 Additives

The physical, mechanical, and optical properties of all previously mentioned groups of plastics can be modified greatly through the admixture of additives. There are many reasons for using additives, such as improving the mechanical properties and cost considerations. Fig. 2.2.8 contains a listing of additives that commonly find application in the processing of plastics. The so-called soluble additives include substances that act chemically, such as stabilizers, for instance, or so-called plasticizers, which are especially important in PVC processing. Substances that act physically, however, such as lubricants, should also be mentioned here. The group of insoluble solids is a completely different group of additives. Pigments and impact modifiers belong to this group. In particular, solids such as mineral powders, carbon black, or wood flour, which make the material harder and also somewhat more economical, are of great importance. At the same time, mention must be made of the variety of fibers–glass, carbon, and aramid fibers—that improve the tensile strength of plastics. Lastly, the insoluble gases that are used as blowing agents to produce foams should be mentioned. A more detailed description of the various groups of plastics and the changes to their properties through the addition of additives is contained in part in the following chapters as well as in the recommended reading at the end of this chapter.

Bibliography for Chapter 2.1

Biederbick, K.H.: Kunststoffe kurz und bündig, Vogel, Würzburg, 1977

Gnauck, B., Fründt, P.: Einstieg in die Kunststoffchemie, 3rd ed., Carl Hanser Verlag, München, Wien, 1991

Rink, G., Schwahn, M.: Einführung in die Kunststoffchemie, Diesterweg, Salle, Sauerländer, 1983

Menges, G.: Werkstoffkunde Kunststoffe, 3rd ed., Carl Hanser Verlag, München, Wien, 1990

N. N.: Kunststoff-Werkstoffe im Gespräch: Aufbau und Eigenschaften, BASF Ludwigshafen, 1975

Batzer, H., Lohse, F.: Einführung in die makromolekulare Chemie, Hüthig & Wepf, Heidelberg, 1976
Vollmert, B.: Grundriß der makromolekularen Chemie, Springer, Heidelberg, 1962
Domininghaus, H.: Kunststoff-Fibel, 3rd ed., Zechner & Hüthig, Speyer, 1980
Elias, H.G.: Makromoleküle, Hüthig & Welp Verlag, Heidelberg, 1981

Bibliography for Chapter 2.2

Biederbick, K.H.: Kunststoffe, kurz und bündig, Vogel, Würzburg, 1977
Gnauck, B., Fründt, P.: Einstieg in die Kunststoffchemie, 3rd ed., Carl Hanser Verlag, München, Wien, 1991
Menges, G.: Werkstoffkunde Kunststoffe, 3rd ed., Carl Hanser Verlag, München, Wien, 1990
Domininghaus, H.: Die Kunststoffe und ihre Eigenschaften, VDI-Verlag, Düsseldorf, 1976
Braun, D.: Erkennen von Kunststoffen, 2nd ed., Carl Hanser Verlag, München, Wien, 1986
Krause, A., Lange, A.: Kunststoff-Bestimmungsmöglichkeiten, Carl Hanser Verlag, München, Wien, 1979
Saechtling, H.: Kunststoff-Bestimmungstafel, Carl Hanser Verlag, München, Wien, 1979
Stoeckhert, K., Woebecken, W.: Kunststoff-Lexikon, 8th ed., Carl Hanser Verlag, München, Wien, 1992
Saechtling, H.: Kunststoff-Taschenbuch, 25th ed., Carl Hanser Verlag, München, Wien, 1992

3 Physical Properties of Plastics

3.1 Thermodynamic Material Properties

To be able to evaluate the processing behavior of plastics, it is necessary to know their thermodynamic and rheological material properties, since initially plastics must usually be in a flowable condition for processing, after which they solidify (thermoplastics) through the removal of heat (cooling) or cure (thermosets, elastomers) through the addition of heat. Because of their molecular structure, the material characteristics of plastics exhibit some very definite differences when compared to other materials, especially the metals. The thermodynamic properties of thermoplastic resins in particular exhibit significant deviations because of the relatively low melting point.

3.1.1 Density

In comparison to other materials, plastics have a quite low *density*. The density range of unfilled plastics extends from about 0.9 g/cm^3 to 2.3 g/cm^3. The best known plastics with a low density are polyethylene (PE) and polypropylene (PP). Both materials are less dense than water. Polytetrafluoroethylene (PTFE) is an example of a polymer with a density exceeding 2 g/cm^3. Fiber-reinforced plastics achieve densities up to approximately 2.3 g/cm^3. If polymers are filled with mineral substances, they may have densities up to 6 g/cm^3, depending on the amount of filler. Most plastics, however, fall in a density range between 1 and 2 g/cm^3.

The density of other materials is often several times higher (aluminum: approximately 2.7 g/cm^3; steel: approximately. 7.8 g/cm^3). The higher density of other materials has two origins. First, the individual atoms (aluminum, iron) are heavier than the carbon, nitrogen, oxygen, or hydrogen atoms. Second, the average distance between the atoms in plastics is sometimes larger than in metals, since plastics consist of macromolecules and not individual atoms.

It must also be recalled that the density of plastics also depends on the respective pressure (p) and temperature (T). In practice, the reciprocal of the density, the specific volume ($v_p = 1/\rho$) is employed to characterize the material. The interrelationships are then usually combined into a so-called *p-v-T* diagram. Fig. 3.1.1 shows examples of diagrams for both an amorphous and a semi-crystalline plastic.

The low density makes plastics an interesting material for all rapidly moving parts (reduction of inertial forces). The possibility of modifying the properties of plastics over wide ranges, however, is a unique aspect of these materials. This is also true for the density. For instance, it is possible and common to foam several plastics, e.g., polystyrene, polyurethane (see Chapter 6.5, "Foaming").

In this case, the volume of the plastic is often increased more than a hundredfold, while the mass remains constant. This means that the density can be reduced to less than 1%. In this

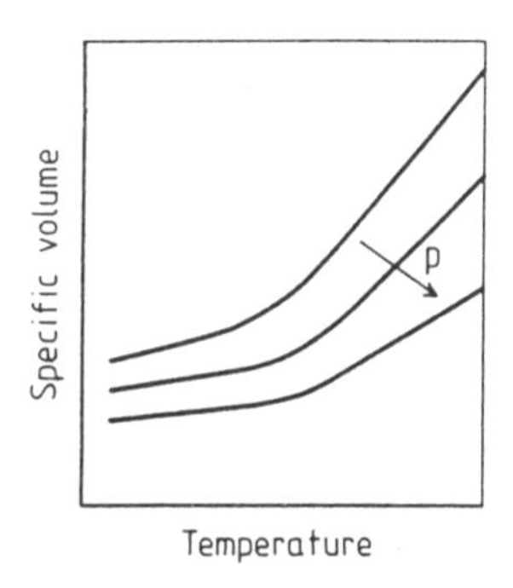

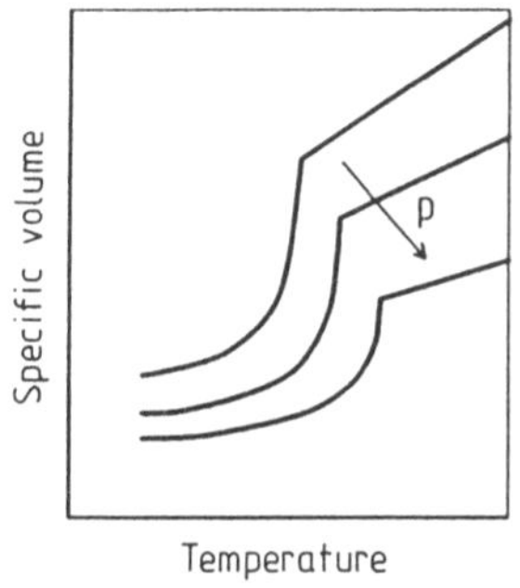

Figure 3.1.1
p-v-T diagrams.

way, foamed plastics often achieve densities below 0.01 g/cm^3. The density of plastics naturally varies with the addition of *fillers* and the like. Even the processing of plastics affects the density, if only to a slight degree. These slight density changes, however, correlate with many other properties (crystallinity, strength, etc.) so that a quick and precise density measurement is very important.

An often used density measuring instrument is the so-called *gradient tube density gauge*. In this measurement method, the test specimen is suspended in a fluid and the density of this fluid is subsequently measured. The following then holds: density of specimen = density of fluid.

3.1.2 Thermal Expansion

Upon heating plastics sometimes expand considerably more than other materials. This is so because the service temperature of plastics is not far below the melting point. Thermoplastics in particular increase their length quite significantly upon heating. The *coefficient of thermal expansion* of polyethylene or polypropylene ($200 \cdot 10^{-6}$ K^{-1}) exceeds that of iron ($12 \cdot 10^{-6}$ K^{-1}) or of aluminum ($24 \cdot 10^{-6}$ K^{-1}) by a factor of 16 or 8, respectively. To offset this disadvantage of plastics when designing precision parts, they are often filled with fibers. In this manner, the thermal expansion is reduced to the value of metals. When filling plastics with carbon fibers, the coefficient of thermal expansion can even become negative. In other words, this material contracts upon heating.

For thermosets, but for some thermoplastics as well (polystyrene, polyvinyl chloride), the coefficients of thermal expansion (70–$100 \cdot 10^{-6}$ K^{-1}) are not quite as high as for polyethylene and polypropylene.

The thermal expansion can be measured as a function of temperature with the aid of *thermomechanical analysis* (TMA). Fig. 3.1.2 illustrates the principle employed in TMA. The oven and the transducer are the heart of this measurement method. The specimen is placed on a ground quartz glass surface and the measurement rod is placed on the specimen. The weight placed on the rod is selected to be as low as possible to avoid any undesirable deformation of the specimen. Every change in thickness resulting from a change in temperature is transmitted to the rod and sensed by the transducer. The result of the measurement is a recording of the signal from the transducer versus the current specimen

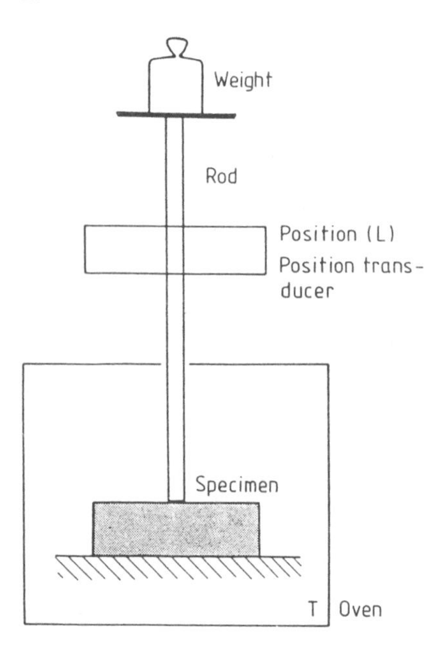

Figure 3.1.2 TMA (thermomechanical analysis).

temperature *T*. This method can be used to determine the glass transition temperature of plastics, since above this value they exhibit a considerably higher coefficient of thermal expansion.

3.1.3 Thermal Conductivity

As do all *insulators*, plastics conduct heat only poorly. In metals, the free electrons provide not only for high electrical conductivity of the material, but also for good thermal conductivity. In plastics, however, which are electrical insulators, these free electrons are absent from the material.

This "insulating" property permits plastics to be used in many applications (e.g., containers for cold and warm liquids), but, on the other hand, it also creates problems with regard to their processing, since the heat required for processing can be introduced and then removed again at the end of processing only slowly.

The thermal conductivity of plastics is about 300 to 1000 times lower than that of metals. Air conducts heat about an additional 10 times more poorly, so that foamed plastics have an even better insulating capacity and are thus employed as *thermal insulating material.* On the other hand, the thermal conductivity of plastics can be increased severalfold by incorporating metallic fillers.

For pure (neat) resins, the thermal conductivity varies over a range of 0.15–0.5 W/mK (Fig. 3.1.3). The temperature has hardly any effect on the thermal conductivity of amorphous thermoplastics. In contrast, the thermal conductivity of semi-crystalline thermoplastics drops with increasing temperature up to the melting point. At the melting point, the conductivity drops a few percent, because the crystalline regions, which conduct heat better

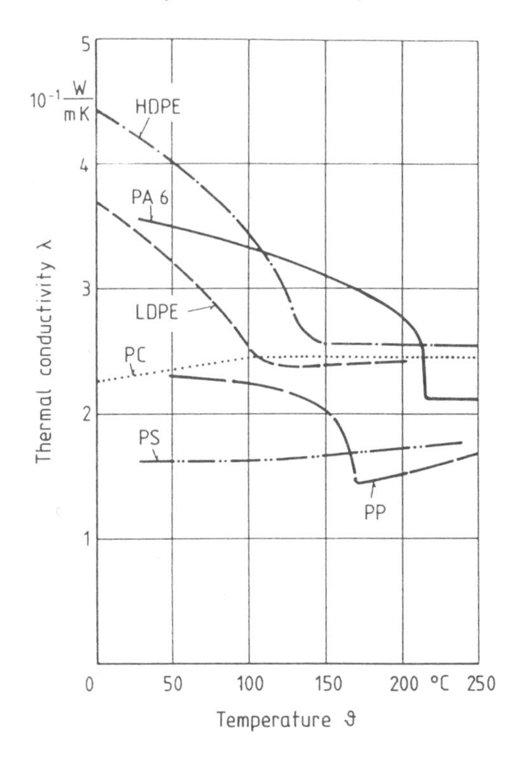

Figure 3.1.3 Thermal conductivity of various plastics.

than the amorphous ones, are molten. Above the melting point, a temperature dependence of the thermal conductivity is no longer present as a rule; in other words, the thermal conductivity of a plastics melt is almost independent of temperature.

3.1.4 Specific Heat Capacity

The *specific heat* of plastics lies in a range of 0.4 (polyurethane) to 2.7 kJ/kg K (Fig. 3.1.4). As is typical for other solids, amorphous thermoplastics exhibit a specific heat capacity that increases continuously with temperature. In contrast, a phase transformation occurs at the crystallite melting temperature in semi-crystalline thermoplastics, i.e., a heat of fusion is needed to melt the individual crystallites. The temperature does not increase during this process. The specific heat, which is defined as the quotient of heat added and temperature increase, thus has a discontinuity at the melting point.

Thermosets change their specific heat capacity quite noticeably when *curing*. During curing heat is released, which leads to a reduction in the specific heat. In a cured thermoset, the specific heat increases monotonically with temperature, since in a cured thermoset neither a chemical reaction nor a physical transformation occurs, even at elevated temperatures.

In the reverse manner, by measuring the specific heat capacity it is possible to obtain information on the type of plastic, amount of filler, or crystallinity. The most commonly employed measurement method is *differential scanning calorimetry* (DSC).

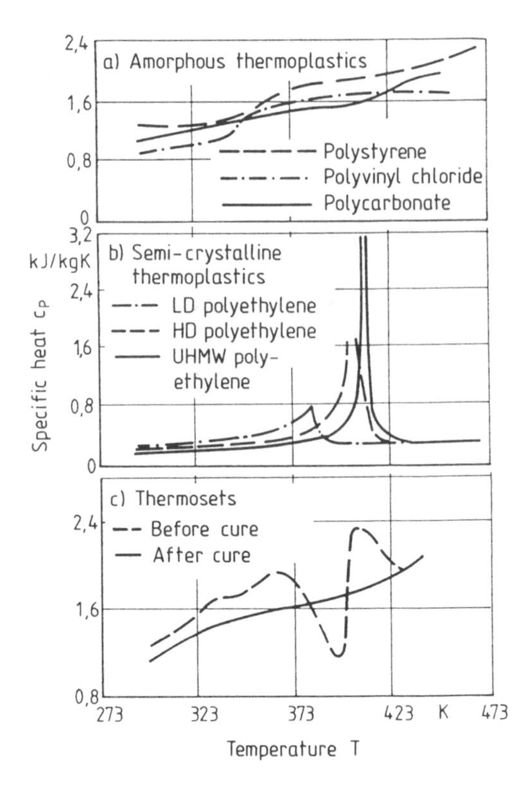

Figure 3.1.4 Specific heat of plastics.

The principle employed in a DSC measurement is shown in Fig. 3.1.5. The specimen to be investigated and a reference specimen are placed in a linearly heated oven (dT/dt = constant). Using a surface thermocouple, a contact temperature that becomes established between the specimen, or the specimen holder, and the oven is measured. This contact temperature is a measure of the heat supplied to the specimen. In comparison to a reference,

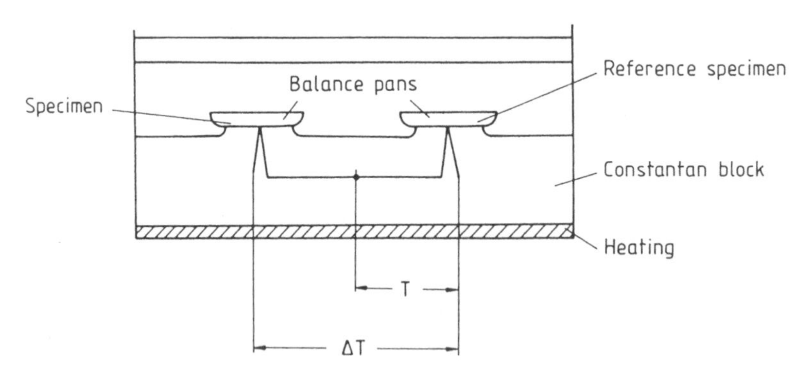

Figure 3.1.5 Principle of a DSC cell.

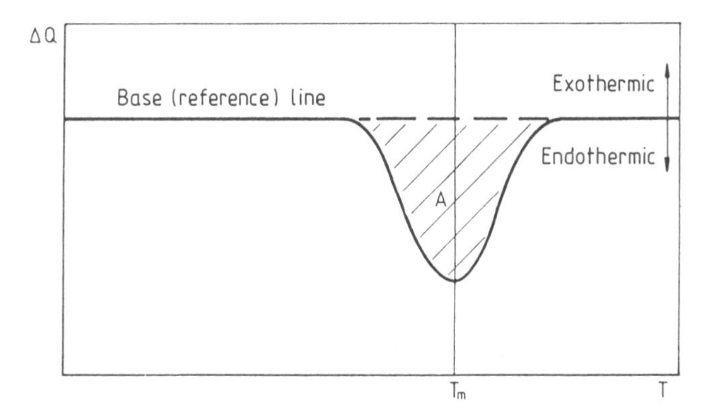

Figure 3.1.6 DSC curve.

for which an empty balance pan may be employed, for instance, the amount of heat supplied to a known amount of material can be quantitatively determined.

The actual advantage of the DSC principle is that calorimetric characteristics of a material can be determined, i.e., heats of reaction, heats of transformation, heat capacities, and the like can be measured in addition to transformation temperatures. By using the instrument constants, the respective temperature difference ΔT measured can be converted into a heat flow ΔQ, which yields the DSC curve shown in Fig. 3.1.6 when plotted versus the respective temperature. This figure shows, in principle, the DSC diagram of a semi- crystalline polymer in the melting range.

From this curve, the melting point T_m of the crystalline regions is obtained as the minimum of the endothermic peak. The area A between the curve and base line is a direct measure of the amount of heat H_m needed to melt the crystallites. The measured amount of heat H_m can now be compared with other values. Thus, the *amount of filler*, for instance, can be determined from the heat values for a filled and an unfilled resin.

3.2 Flow Properties of Polymer Melts

Describing, explaining, and measuring the flow properties of materials—such as those of plastics—is the primary objective of the "science of the deformation and flow of bodies," which is called *rheology*. Rheometry, the technology of measuring the flow behavior, is a subfield of rheology.

In plastics processing, the materials to be processed must usually be in a flowable condition. This can be accomplished through an increase in temperature with resultant melting, by dissolving the material in a solvent or even the direct processing of low-molecular-weight products that subsequently react to completion. During such processing, the *viscosity*, which is a measure of the internal resistance of the material to a constantly applied force during flow, is the property of interest.

During the flow process as it occurs in plastics processing machinery, the melt is subjected primarily to shear. This so-called shear flow results because, as a rule, plastics melts adhere to the surfaces of the tooling being used to shape them (Stoke's adhesion, wall adhesion). This can be illustrated in simplified form quite well with the two-plate model (Fig. 3.2.1).

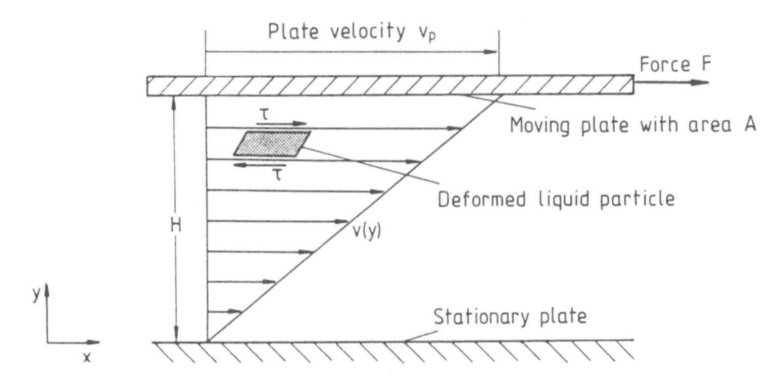

Figure 3.2.1 Schematic illustration of shear flow under laminar flow conditions.

If one plate is displaced with respect to the other, the layers of liquid slide past one another, and the melt is sheared. An element of fluid is deformed as shown in Fig. 3.2.1. The result is simple *shear flow* with the *shear rate*

$$\dot{\gamma} = \frac{dv}{dy} = \frac{v_p}{H} \tag{1}$$

and the shear stress

$$\tau = F/A \tag{2}$$

In this simplest type of shear flow, the shear stress and shear rate are constant, and the flow rate $v(y)$ increases linearly with height y.

3.2.1 Newtonian and Non-Newtonian Fluids

Under stationary shear flow, there is a shear stress between the fluid layers. In the simplest case of a Newtonian fluid, the shear stress is proportional to the shear rate. Thus,

$$\tau = \eta \cdot \dot{\gamma} \tag{3}$$

The proportionality factor η is called the dynamic *shear viscosity*, or only the *viscosity* for short. It has the dimension Pa s (Pascal seconds).

The relationship between shear stress and shear rate (or, more generally, between stress and deformation rate) as given in Eq. (3) is called the *flow law* or *rheological equation of state.*

Equation (3) also represents the definition of a Newtonian fluid in which the viscosity η is constant. The best known representative of this simplest class of fluids is water (viscosity: 10^{-3} Pa s), but alcohols, gasolines, and many oils also exhibit Newtonian behavior.

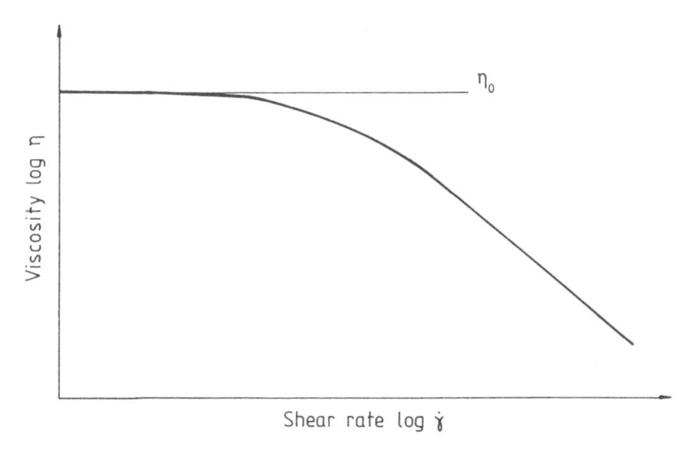

Figure 3.2.2 General behavior of viscosity as a function of shear rate.

In contrast, plastics melts generally exhibit non-Newtonian behavior. Their viscosity is not constant, but rather a function of the shear rate. In analogy to the relationship valid for Newtonian fluids, the flow law is written:

$$\tau = \eta(\dot{\gamma}) \cdot \dot{\gamma} \tag{4}$$

or

$$\eta(\dot{\gamma}) = \tau / \dot{\gamma} \neq \text{const} \tag{5}$$

Liquids that behave in accordance with Eqs. (4) and (5) are called *non-Newtonian fluids.*

If the viscosity η is plotted as a function of the shear rate in a log–log diagram, the general curve shown in Fig. 3.2.2 is obtained, as a rule, for polymers (at constant temperature). It is seen that for low shear rates the viscosity remains constant, i.e., the flow is Newtonian in this region, but that with increasing shear rate, however, the viscosity starts to decrease continuously. This behavior is also called *pseudoplastic*. The constant viscosity at low shear rates is called the *lower Newtonian limit* or the *zero-shear viscosity*.

In addition to the plot of the viscosity versus the shear rate in the so-called *viscosity curve*, illustration of the relationship between the shear stress and the shear rate (also in a log–log plot) as the *flow curve* is common (Fig. 3.2.3).

In a Newtonian fluid the shear stress is proportional to the shear rate. The log–log plot of the values thus yields a curve with a slope of 1, i.e., the angle between the abscissa and the flow curve is 45°.

For a pseudoplastic fluid, curves are obtained with a slope greater than 1, i.e., the shear rate increases progressively with increasing shear stress. By the same token, this means that the shear stress increases only degressively with the shear rate.

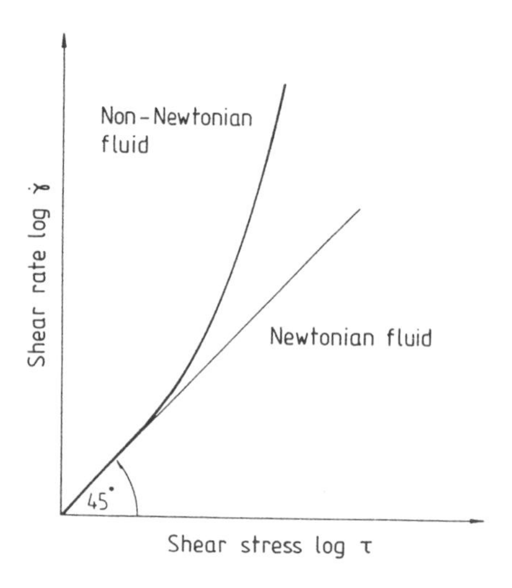

Figure 3.2.3 General behavior of the shear rate as a function of the shear stress.

3.2.2 Laws for Describing Non-Newtonian Melt Behavior

Ostwald and de Waele Power Law

If the flow curves of different polymers are presented in a log–log plot, curves are obtained that consist of two approximately linear sections and a transition region (Fig. 3.2.4). In many cases, one is operating in only one of the two regions, so that to mathematically describe this section of the curve a function of the form

$$\dot{\gamma} = \Phi \cdot \tau^{m} \tag{6}$$

is adequate. Equation (6) represents the so-called *Ostwald and de Waele Power Law* with the two parameters flow exponent m and fluidity Φ. The flow exponent m characterizes the flow properties of a material and its deviation from Newtonian behavior. It follows that:

$$m = \frac{\Delta(\log\dot{\gamma})}{\Delta(\log\tau)} \tag{7}$$

where m is the slope of the flow curve in the region of interest when plotted in a log–log diagram (Fig. 3.2.4).

As a rule, the value of m lies between 1 and 6 for plastics melts and, in the shear rate range of approximately 10^0–10^4 s^{-1} that is relevant for the design of screws and tooling for plastics processing, usually only between 2 and 4. For $m = 1$, $\Phi = 1/\eta$, i.e., the fluid exhibits Newtonian flow behavior.

With

$$\eta = \tau/\dot{\gamma}$$

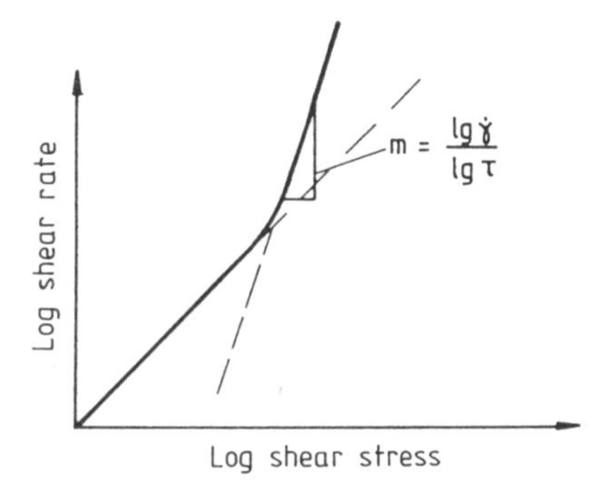

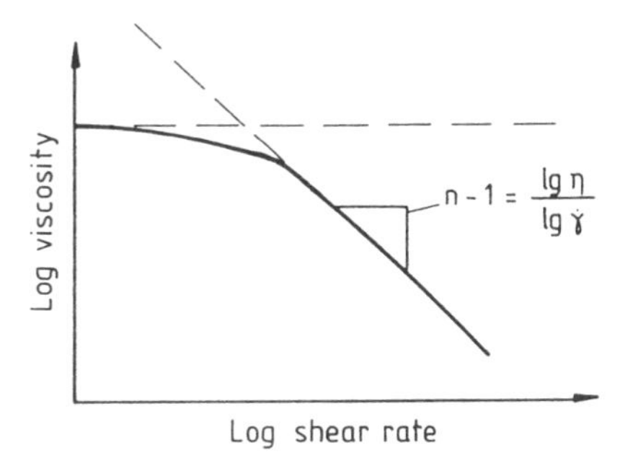

Figure 3.2.4 Description of the flow and velocity curves with the aid of the power law.

the following is obtained from Eq. (6) for the viscosity function:

$$\eta = \Phi^{-1} \cdot \tau^{1-m} \tag{8}$$

$$= \Phi^{-1/m} \cdot \dot{\gamma}_{1/m-1}$$

With

$$k = \Phi^{-1/m}$$

and

$$n = \frac{1}{m}$$

the following is obtained for the usual representation of the viscosity function:

$$\eta = k \cdot \dot{\gamma}^{n-1} \tag{9}$$

The factor k is called the consistency factor. It represents the viscosity at a shear rate of $\gamma = 1\ \mathrm{s}^{-1}$. The viscosity exponent n is equal to 1 for Newtonian behavior and is <1 for most polymers. It describes the slope of the viscosity curve (Fig. 3.2.4).

The empirically determined power law is often employed in practice, since it is quite easy to handle mathematically. A disadvantage, however, is that it describes only sections of flow and viscosity curves well. Furthermore, for mathematical reasons, the viscosity approaches

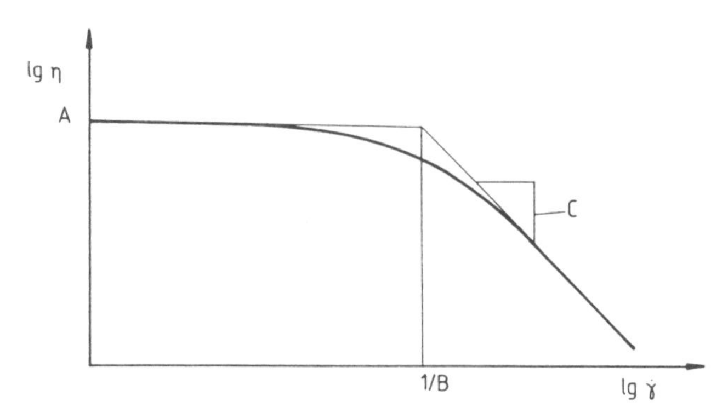

Figure 3.2.5 Description of a viscosity curve with the aid of the Carreau model.

infinity when the shear rate approaches zero. This means that the approximately shear-rate-independent Newtonian region cannot be described.

Carreau Law

The *Carreau Law* describes the viscosity curve by means of a three-parameter equation of the form

$$\eta = \frac{A}{(1 + B\dot{\gamma})^{c}} \tag{10}$$

where A represents the zero-shear viscosity, B the so-called reciprocal transition velocity, and C the slope of the viscosity curve in the non-Newtonian region (Fig. 3.2.5).

The advantage of this model is that it correctly represents the actual material behavior over a broader range of shear rates than does the power law and also yields meaningful viscosity values for $\gamma \to 0$. The coefficients, B, and C can only be determined from measured data by means of regression analysis. Thus, more effort is involved than with the power law, where the parameters can easily be determined graphically from a line giving the best fit.

3.2.3 Effect of Temperature on Flow Behavior

The flow behavior of polymer melts depends not only on the shear rate, but also on the temperature T and (to a slight degree) on the hydrostatic pressure p. (Because of its usually minimal importance in actual practice, the pressure dependence of the viscosity shall not be discussed.)

If for one and the same polymer melt the viscosity curves for different temperatures are presented in a log–log plot (Fig. 3.2.6), it is seen that, although the location of the curves in the plot changes with temperature, the shape of the viscosity curve remains the same, as a rule. In a log–log plot as shown in Fig. 3.2.6, the curves can be superimposed on one another if they are shifted along a line with a slope of -1 (*temperature shift principle*). All

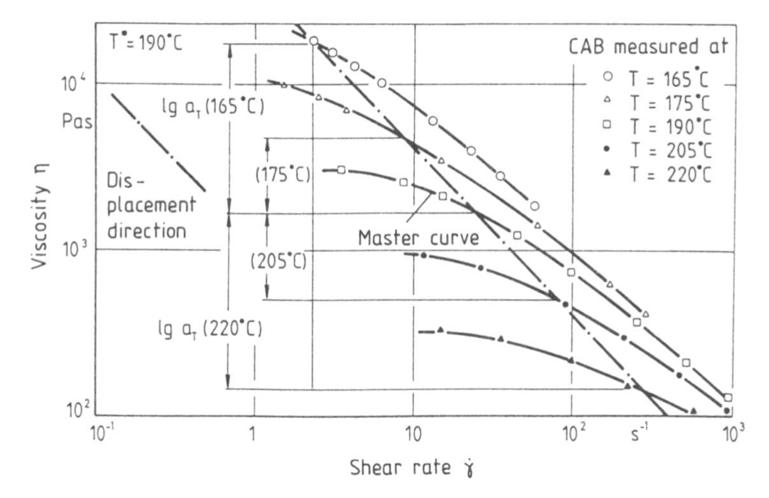

Figure 3.2.6 Viscosity functions at various temperatures (shift principle: temperature shift factor a_T [T]).

of the curves fall on a so-called *master curve.* This master curve is independent of the temperature, i.e., it is temperature-invariant (Fig. 3.2.7).

Mathematically, the temperature-invariant plot is obtained by displacing the individual viscosity curves simultaneously downward and to the right by the amount log $\eta > 0\,(T)$ in accordance with the respective zero-shear viscosity $\eta > 0\,(T)$. This displacement yields a plot of the so-called reduced viscosity $\eta/\eta > 0$ versus the quantity $\eta > 0 \cdot \gamma$. A single characteristic function thus results for the polymer:

$$\frac{\eta(\dot{\gamma}, T)}{\eta_0(T)} = f(\eta_0(T) \cdot \dot{\gamma}) \tag{11}$$

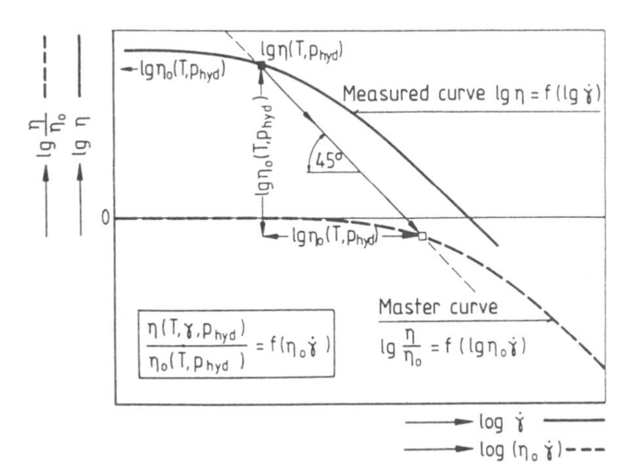

Figure 3.2.7 Temperature- and pressure-invariant viscosity functions.

where the temperature T is freely selectable as a reference variable. If the viscosity function for a certain temperature is sought, and only the master curve or the viscosity curve for another temperature is available, the temperature must be shifted in order to obtain the desired curve. The amount by which this viscosity curve must be displaced, however, is not known initially. Various models for determining this so-called temperature shift factor a_T are known. Of these, the quite universal law of Williams, Landel, and Ferry (*WLF Law*) is presented here by way of example. According to this law:

$$\log a_T = \log \frac{\eta(T)}{\eta(T_S)} = -\frac{8{,}86(T - T_S)}{101{,}6 + (T - T_S)} \tag{12}$$

where a_T is the so-called temperature shift factor and T_S is the so-called standard temperature. Because

$$\log \frac{\eta(T)}{\eta(T_s)} = \log \eta(T) - \log \eta(T_S) \tag{13}$$

$\log a_T$ represents the amount by which the viscosity curve η (T_S) must be shifted along the 45° line toward the ordinate in a log–log plot (Fig. 3.2.6)

The standard temperature T_S is material-dependent. It lies about 50°C above the softening temperature (glass transition temperature T_G) at p = 1 bar (14.5 psi).

3.2.4 Measurement of Viscous Flow Properties

The most important material property used to describe the flow behavior is the dynamic viscosity η. If the measurement is limited to this quantity, the corresponding measuring instrument is called a *viscometer*. In contrast, *rheometers* also permit the determination of other, largely elastic, quantities. The following discussion will be limited initially to the measurement of the dynamic viscosity, i.e., to viscometers and the field of viscosimetry.

3.2.4.1 Capillary Viscometers

Capillary viscometers are the most commonly employed viscometers. The reasons for this can be found in the relatively simple handling and evaluation of the measurements, together with the possibility of achieving high shear rates that are relevant to actual practice. In addition, certain capillary viscometers can be operated in a bypass arrangement as a continuous monitoring device directly on the processing machine, e.g., extruder.

Construction and Measurement Principle

The measurement principle utilized often in laboratory units is based on forcing the melt of interest through a capillary with a circular cross section (annular or slot-shaped cross sections are also possible) by means of a plunger. Usually, the plunger speed, and thus the volumetric flow rate, are fixed, and the pressure drop in the capillary is measured with the aid of one or several pressure sensors. In the case of a circular capillary, the pressure sensor

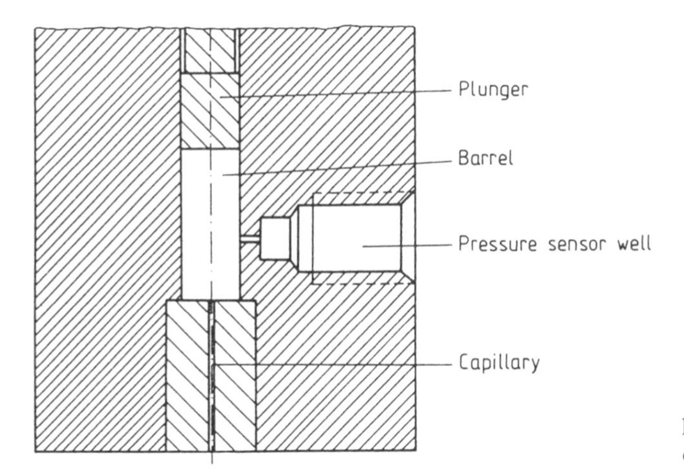

Figure 3.2.8 Principle of a capillary viscometer.

is not installed in the capillary itself because of the small bore diameter (usually 1–3 mm), but rather in the considerably larger plunger cylinder. The ambient pressure at the capillary outlet serves as a second reference point (Fig. 3.2.8).

For Newtonian fluids, the flow in the circular capillary is described by the *Hagen–Poiseuille Law*. It is assumed that the flow is laminar, stationary, isothermal, and adheres to the wall. According to this law, the following results for the shear stress at the capillary wall ($r = R$):

$$\tau_W = \frac{\Delta p \cdot R}{2L} \tag{14}$$

The shear rate at the wall is given by

$$\dot{\gamma}_W = \frac{4\dot{V}}{\pi R^3} \tag{15}$$

With

$$\eta = \frac{\tau_W}{\dot{\gamma}_W}$$

one point (η, γ) on the viscosity curve can be established from known geometric quantities (R, L), the specified volumetric flow rate V, and the measured pressure difference. By varying V, and thus γ, the behavior can be determined over a wide range of shear rates.

Methods to Correct for Non-Newtonian Melt Behavior

For all of the equations derived above it was assumed that the fluid exhibits Newtonian behavior. Since, however, this is generally not the case with polymers, Eqs. (14) and (15) yield a false, i.e., a so-called *apparent, flow curve*, which represents the shear stresses as a function of the apparent shear rate D_S.

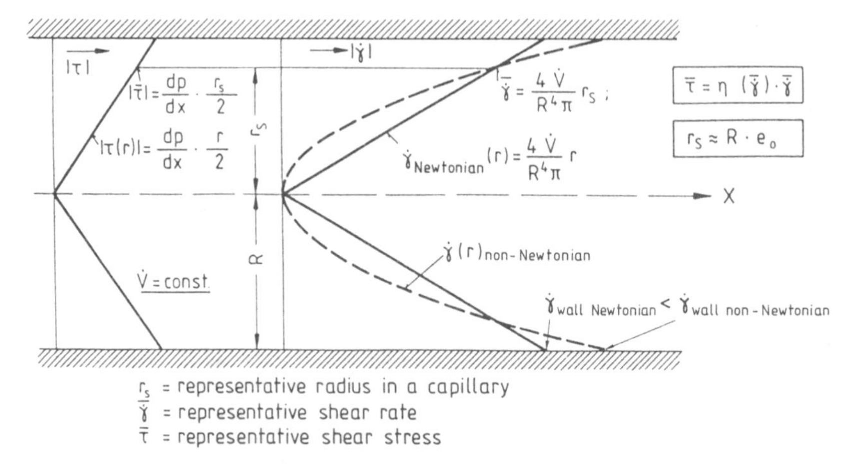

Figure 3.2.9 Determination of the viscosity from the "representative" shear rate and shear stress in a capillary rheometer (circular cross section).

To obtain the *true viscosity curve*, which takes into consideration the non-Newtonian behavior, two correction methods are commonly employed. With the *Weissenberg–Rabinowitsch correction*, the true shear rate is obtained by differentiating the apparent flow curve. Division of the shear stress values from Eq. (14) by the true shear rates then yields the true viscosities. This method involves quite a bit of effort because of the point-by-point graphical differentiation of the flow curve.

The *concept of representative viscosity*, first described by Schümmer and developed further by Giekus and Langer, yields the same accuracy with considerably reduced effort. This correction method is based on the fact that for a laminar and isothermal pressure flow there is one location in the flow channel where the shear rates of a Newtonian and a non-Newtonian fluid are identical if the same volumetric flow rate V is assumed. Determining the shear stress and the shear rate for a Newtonian fluid at this location yields the true viscosity for the "representative" shear rate or shear stress (Fig. 3.2.9).

Assuming that the behavior of the fluid can be described by the power law, the location of the intersection—i.e., the representative distance from the center of the flow channel—can be calculated (Fig. 3.2.10). In principle, this location depends on the flow behavior of the fluid, but, for the flow exponent range ($2\, m \leq 4$) of interest for actual practice, can with sufficient accuracy be considered constant. These representative distances e from the center of the flow channel are obtained as

$$e_0 = \frac{r_s}{R} = 0{,}815$$

for a round pipe and as

$$e_\square = \frac{h_s}{H/2} = 0{,}772$$

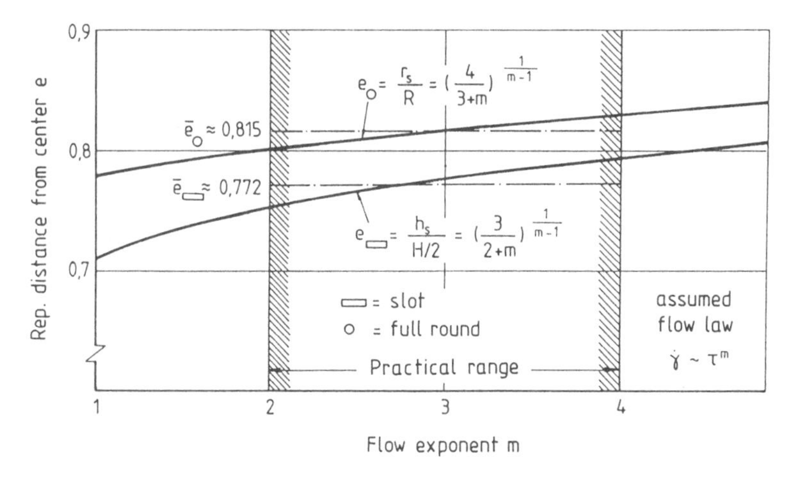

Figure 3.2.10 Representative distance from the channel center for plastics melts with different flow exponents.

for a flat slot. The resulting errors are very small (<1.8% for flow through a round pipe: 2,5% for flow through a slot). If the apparent shear rate D_s at the flow channel wall is known from experiments, the representative shear rate γ is obtained as:

$$\dot{\gamma} = D_s \cdot e \tag{16}$$

The true viscosity—and representative viscosity—are calculated as

$$\eta = \frac{\tau_W}{D_s} = \frac{\tau_W \cdot e}{\dot{\gamma}} \tag{17}$$

at the location of the representative shear rate.

Consideration of Elastic Inlet Effects

In capillary viscometers the pressure before the capillary is often measured (Fig. 3.2.8) instead of the pressure drop between two pressure sensors in the capillary. In this case, not only the pressure drop resulting from frictional forces of the fluid in the capillary is sensed (only this may be utilized in the corresponding equations for calculating viscosity), but also a so-called *inlet pressure loss* associated with elastic deformation of the melt as it enters the capillary. Where the flow channel contracts, the melt particles are stretched (elongated) severely in the flow direction. This deformation is transported to some extent through the capillary in the form of elastically stored energy and is regained at the outlet in the form of swelling of the melt strand. This elastically stored energy must first be supplied in the form of additional pressure. This inlet pressure loss is also sensed by the pressure sensor and thus must be deducted from the result. This is accomplished by the so-called *Bagley correction.*

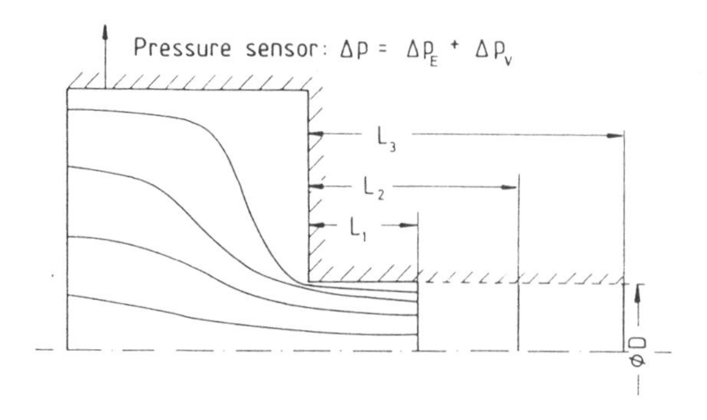

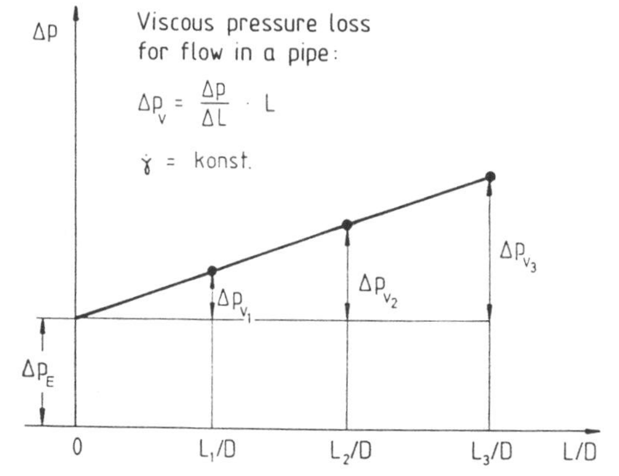

Figure 3.2.11 Bagley correction.

The Bagley correction (after E.B. Bagley) is based on two assumptions:

1. The inlet pressure loss occurs only at the inlet to the capillary and is thus independent of the capillary length.
2. Within the capillary, the pressure drops with a constant pressure gradient because of viscous flow.

If the pressure drop across capillaries with different L/D ratios is recorded at constant shear rate (usually the diameter D is kept constant and the length L is varied), the results fall on a line in a p-L/D diagram (Fig. 3.2.11). The slope of this line corresponds to the pressure gradient in the capillary. By varying the shear rate, a family of lines is obtained with the shear rate as a parameter.

Extrapolating these lines to a nozzle length of zero ($L/D = 0$) yields a value on the ordinate corresponding to the length-independent inlet pressure loss. This inlet pressure loss increases with increasing shear rate, just like the pressure gradient across the capillary. To calculate the viscosity curve, the difference between the measured pressure p_{meas} and the inlet pressure loss p_i must be calculated:

$$\Delta p = \Delta p_{gem} - \Delta p_e \tag{18}$$

Only this difference can be used in Eq. (14) for the shear stress.

3.2.4.2 *Rotational Viscometers*

In capillary viscometers there is a pressure flow with a shear rate that varies over the cross section. In rotational viscometers, on the other hand, a drag flow is generated with a shear rate that, under ideal conditions, is constant over the cross section. As a result, none of the corrections needed for capillary viscometers are necessary.

The shear stress is determined from the torque required for rotation, while the shear rate is calculated from the angular velocity. The viscosity is then obtained by dividing the shear stress by the shear rate.

Rotational viscometers are preferred with low and moderate shear rates. Typical rotational viscometers are the Couette and the cone-and-plate viscometer.

3.3 Elastic Properties of Polymer Melts

Plastics melts do not exhibit only viscous behavior, but also possess an elasticity that is not inconsiderable. Their properties lie between those of an ideal fluid and those of an ideal (Hookian) solid, and they are thus said to exhibit *viscoelastic behavior* or *viscoelasticity*. The viscoelastic behavior has a decisive influence on the processing and molded part properties of a plastics resin.

The reasons for the viscoelasticity of melts can be found in the molecular structure of plastics, which is why the structure directly effects the properties. The long chain molecules, which often have numerous branches, do not slip past one another as rapidly during flow as do compact water molecules, for instance. Entanglements form that hinder flow.

In addition, every type of flow aligns the normally randomly arranged molecules to a greater or lesser degree so that they assume a kind of stretched orientation. Opposing this is the tendency of the molecules to reassume a condition with the greatest possible disorder (state of highest entropy). These microscopic processes lead to the following macroscopic behaviors that are completely unknown in purely viscous fluids.

1) Normal Stresses

In plastics melts, normal stresses that differ in magnitude form in all three spatial directions under strictly shear flow conditions. This property of plastics melts is called the *normal stress effect*. The differences between the normal stresses can be several times larger than the shear stress.

2) Time-Dependent Behavior

If a plastics melt is suddenly subjected to a constant shear rate, the shear stress increases only gradually. If the shearing action is stopped abruptly after a certain period of time, it takes a while for the stress to decay (*relaxation*); this behavior can also be demonstrated in a tensile test. If, on the other hand, a plastics melt is deformed and the deforming force is suddenly removed, the melt will reduce some of the initial deformation by attempting to return to the undeformed condition (*retardation*).

3.4 Cooling from the Melt

In the melt and in solution the individual macromolecules— considered as threads—of a thermoplastic exhibit a coiled structure similar to that of cotton wadding or spaghetti, since this state has the maximum entropy. The entanglements of the individual molecules with one another open and close as a result of the thermal motion in the melt, with the result that, on average, a random arrangement is retained. If a thermoplastic melt is cooled, the chains increasingly lose their mobility and the melt becomes highly viscous.

3.4.1 Solidification of Amorphous Thermoplastics

If the polymer involved is an amorphous material, the increasingly dense packing associated with cooling does not lead to crystallization, but rather the melt continues to solidify in a disordered state until a glasslike condition is reached below the range of the solidification temperature. Passage through the glass transition temperature is characterized by pronounced changes in the physical and mechanical properties (see also Chapter 2).

3.4.2 Solidification of Semi-Crystalline Thermoplastics

Macromolecules that exhibit irregularity in the molecular chain can form crystalline structures during cooling. Since, however, complete crystallization cannot occur because of the size of the molecules, such materials are called *semi*-crystalline polymers. The *degree of crystallization* of commercially available polymers lies between 20 and 80%. When cooling from the melt, a spherulite consisting of crystalline and amorphous regions grows around a crystallization nucleus in semi-crystalline polymers. A spherulite grows spherically into the amorphous surroundings until it collides with another spherulite. The spherulites can be easily recognized when viewed in polarized light under a microscope with crossed polarizing filters (Fig. 3.4.1).

Several properties of semi-crystalline materials depend largely on the size of the spherulites and the degree of crystallization. The dependence of the modulus of elasticity on the degree

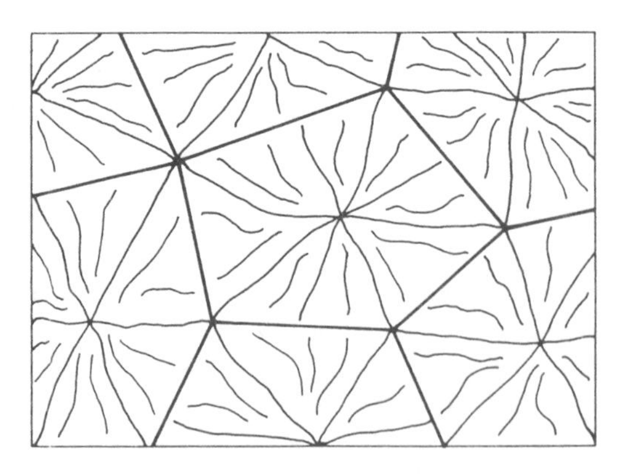

Figure 3.4.1 Illustration of a coarse spherulitic polypropylene structure.

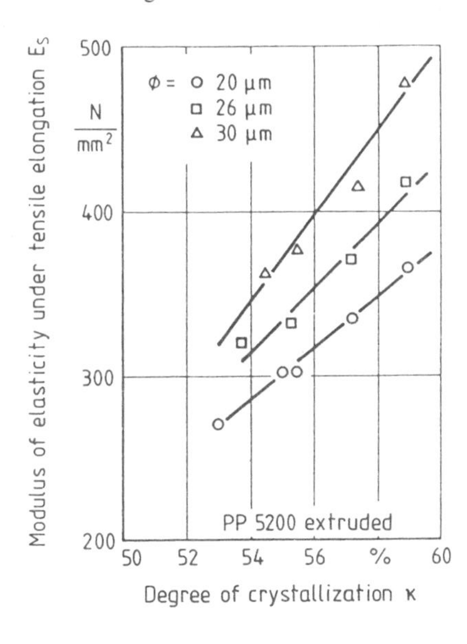

Figure 3.4.2 Modulus of elasticity under tensile elongation as a function of the internal properties.

of crystallization and spherulite diameter for a polypropylene is mentioned here as one example. This relationship is shown in Fig. 3.4.2.

It is obvious that knowledge of the formation of spherulites and ways of affecting their number and size, for instance, are important for optimum processing of *semi-crystalline thermoplastics.*

Crystallization occurs in three stages:

- nucleation,
- crystal growth (primary crystallization),
- recrystallization (secondary crystallization).

The development of spherulites begins with the formation of crystallization nuclei. A nucleus capable of growth grows until the resulting spherulite collides with another spherulite; this concludes the primary crystallization phase. Recrystallization already begins during crystal growth behind the advancing spherulite front. Some of the amorphous material contained in the spherulite crystallizes, with a resulting decrease in volume.

The transformation rate during crystallization is determined by the number of nuclei and the growth rate. Since crystallization is an exothermic process, it can be followed with calorimetric measurement methods. For instance, it is possible to follow crystallization at a constant cooling rate with the aid of *DSC analysis* (see Section 3.1). The temperature with the highest transformation rate is called the *crystallization temperature* T_k.

As shown in Fig. 3.4.3, the crystallization temperature is the temperature at which the exothermic peak has its maximum. As a melt cools, the number of thermal and athermal nuclei increases with increasing undercooling of the melt. The cooling rate thus has a great effect on the kinetics of crystallization and formation of the microstructure.

At high cooling rates, the crystallization temperature is shifted to lower values, and the many spherulites form a fine structure. If cooling is slower, few nuclei form, and the resulting

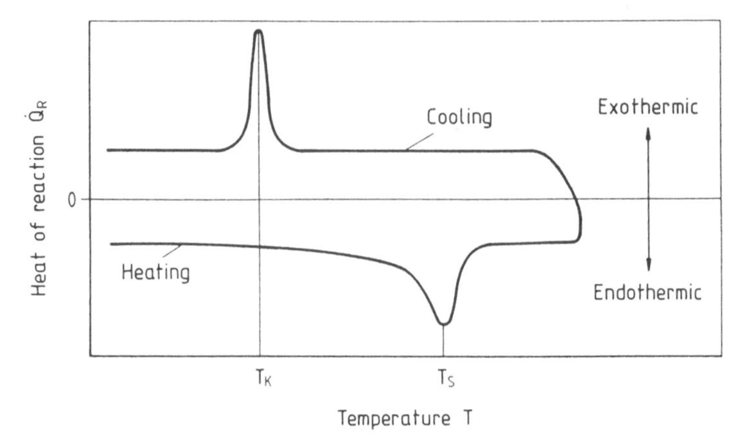

Figure 3.4.3 Heating and cooling steps during the DSC analysis of polypropylene.

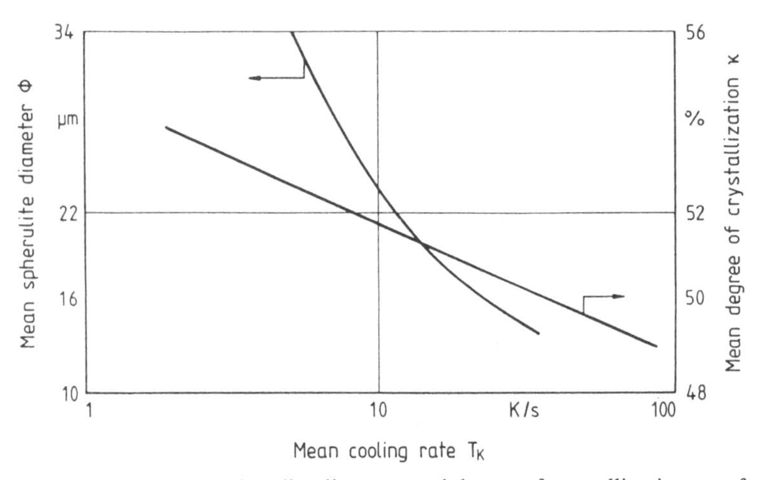

Figure 3.4.4 Mean spherulite diameter and degree of crystallization as a function of the mean cooling rate.

spherulites are larger. The above-mentioned relationship between microstructure and cooling rate is illustrated by the curves shown in Fig. 3.4.4. The decrease in mean spherulite diameter with increasing cooling rate for a polypropylene can be clearly recognized. At the same time, the degree of crystallization drops, since the crystallization rate decreases with decreasing temperature.

3.4.3 Nucleation

The growth rate of the spherulites—and thus the crystallization rate—is highly temperature-dependent, as was already shown. Individual plastics also exhibit significant differences in

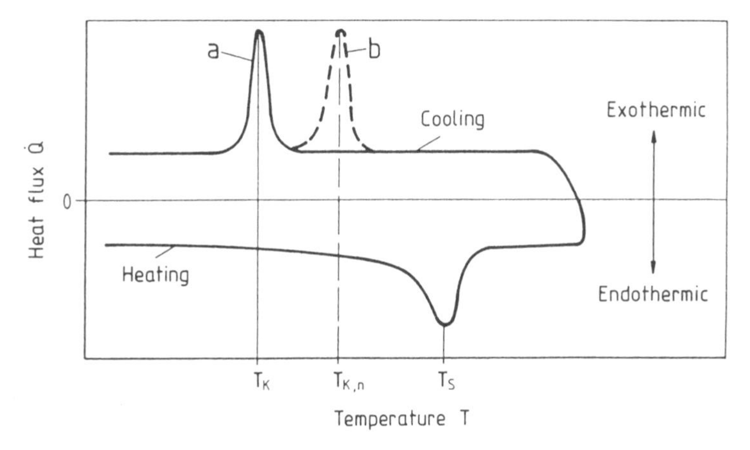

Figure 3.4.5 Recording of a DSC curve for analysis of polypropylene (of base material; b: nucleated material).

their crystallization behavior. The following table lists the maximum spherulite growth rates for several semi-crystalline thermoplastics.

Maximum crystallization rate (μm/s)

Polyethylene	33
Polypropylene	$3.3\ 10^{-1}$
Polycarbonate	$1.7\ 10^{-4}$
Polyamide 6	3.3
Polyamide 66	20

As a rule, a fine microstructure with the highest possible degree of crystallization is desirable immediately after part formation, but this cannot always be accomplished solely through appropriate processing parameters. With slowly crystallizing polymers in particular, for instance, polypropylene, these two requirements are in opposition to one another.

To achieve a fine microstructure in spite of slow cooling, even with slowly crystallizing thermoplastics, so-called nucleating agents are added to increase the number of heterogeneous nuclei in the melt. *Nucleation* results in earlier and more rapid crystallization, which is thus also completed sooner. A fine-grained microstructure with a narrow spherulite size distribution results even with slow cooling. To evaluate the effectiveness of nucleation, the crystallization temperature, as measured with the aid of DSC analysis at a constant cooling rate, for instance, can be utilized. By way of example, Fig. 3.4.5 shows a polypropylene both with and without nucleation. Shifting of the crystallization temperature to a higher value is clearly noticeable for the nucleated material.

3.4.4 Determining the Degree of Crystallization

Measurement of a specimen's *density* is the most important and simplest way of determining the degree of crystallization. This method utilizes a two-phase model for the semi- crystalline material (see also Chapter 2). It is assumed that both a purely crystalline and a purely amorphous phase are present, with the actual density determined by the volume fractions of the two phases.

The following equation relates the density ρ and the degree of crystallization κ:

$$\varkappa = \frac{\rho - \rho_a}{\rho_c - \rho_a}$$

With unfilled specimens, which cannot have any voids or entrapped air, this method permits exact determination of the degree of crystallization if the density values for the purely amorphous and purely crystalline materials are known. In addition to this relatively simple method of measurement, there are a number of other methods for determining the degree of crystallization, some of which are quite involved, but they will not be discussed further here. Instead, the reader is referred to the appropriate technical literature.

Bibliography for Chapter 3.1

Menges, G.: Werkstoffkunde Kunststoffe, 3rd ed., Carl Hanser Verlag, München, Wien, 1990
Domininghaus, H.: Die Kunststoffe und ihre Eigenschaften, VDI-Verlag, Düsseldorf, 1986
Hemminger, W.F., Commenga, H.K.: Methoden der Thermischen Analyse, Springer Verlag, Berlin, 1989

Bibliography for Chapters 3.2 and 3.3

Böhme, G.: Strömungsmechanik nicht-newtonscher Fluide, Stuttgart: Teubner, 1981
Ebert, F.: Strömung nicht-newtonscher Medien, Braunschweig, Wiesbaden: Vieweg, 1980
Giesekus, H., Langer, G.: Die Bestimmung der wahren Fließkurven nicht-newtonscher Flüssigkeiten und plastischer Stoffe mit der Methode der repräsentativen Viskosität, Rheol. Acta 16 (1977), p. 1–22
Menges, G.: Werkstoffkunde Kunststoffe, 3rd ed., Carl Hanser Verlag, München, Wien, 1990
Michaeli, W.: Extrusionswerkzeuge für Kunststoffe und Kautschuk: Bauarten, Gestaltung und Berechnungsmöglichkeiten, 2nd ed., Carl Hanser Verlag, München, Wien, 1991
Plajer, O.: Praktische Rheologie für Kunststoffschmelzen, Speyer: Zechner & Hüthig, 1970
VDI (Hrsg.): Praktische Rheologie für Kunststoffe, Düsseldorf: VDI-Verlag, 1983
VDI (Hrsg.): Praktische Rheologie der Kunststoffschmelzen und Lösungen, Düsseldorf: VDI-Verlag, 1983
Williams, M.L., Landel, R. F., Ferry, J.D.: The Temperatur Dependence of Relaxation Mechanism in Amorphous Polymers and other Glass-forming Liquids, J. Am. Chem. Soc. 77 (1955) 7, p. 3701–3706
Schümmer, P., Worthoff, R.H.: An elementary method for the evaluation of a flow curve, Chem. Eng. Sci. 38 (1978), 759–763
Bagley, E.B.: End corrections in the capillary flow of polyethylene, J. Appl. Phys. 35 (1961) 9, p. 2767–2775

Bibliography for Chapter 3.4

Van Krevelen, D.W.: Properties of Polymers Elsevier Scientific Publishing Company, Amsterdam–Oxford–New York
Menges, G.: Werkstoffkunde Kunststoffe, 3rd ed., Carl Hanser Verlag, München, Wien, 1990

Kämpf, G.: Charakterisierung von Kunststoffen mit physikalischen Methoden, Carl Hanser Verlag, München, Wien, 1982
Dominghaus, H.: Die Kunststoffe und ihre Eigenschaften, VDI-Verlag Düsseldorf, 2nd ed., 1986
Hoffman, Krömer, Kuhn: Polymeranalytik I, II, Thieme Verlag, Stuttgart
Saechtling, H.: Kunststoff-Taschenbuch, 25th ed., Carl Hanser Verlag, München, Wien, 1992
N.N.: Reihe Ingenieurwesen, Kühlen von Extrudaten, VDI-Verlag, Düsseldorf, 1978
Elias, H.: Quantitative Methoden der Morphologie, Springer Verlag, Berlin, Heidelberg, New York, 1967

4 Materials Science of Plastics

4.1 General Behavior of Materials

The development of plastics led to the expression "tailor-made materials." Plastics have been optimized on the basis of market requirements, with a view to improved *processibility* and by providing specific *property profiles*. The continual expansion of the spectrum of materials and applications is the result primarily of *modification* of resins through incorporation of *additives* (as reinforcements, fillers, plasticizers), *mixing* (blending) with other polymers, and the development of new polymers and so-called composites.

The technical and technological behavior of plastics is determined largely by the chemical nature of the basic compounds, the length and arrangement of the macromolecules. Depending on the molecular structure, all states of matter between solid and liquid can be present.

The temperature dependence of the *shear modulus* and of the *mechanical damping* can be employed to classify and to characterize the mechanical-thermal behavior of plastics. In the application and testing of plastics, the structure-related effects of temperature and time (as well as load duration and rate of loading) on the deformation behavior have to be borne in mind (Fig. 4.1.1). The following fundamental properties are generally valid with regard to the *deformation behavior* of plastics:

- Plastics are *viscoelastic* materials (Fig. 4.1.2); this means that the behavior of the material in response to an external load is a function of time. Under constant load, the deformation increases with time (*creep retardation*). Under constant deformation, the stress decreases with time (*recovery*, *relaxation*).
- Plastics are *non-linear viscoelastic* materials; this means that the magnitude of the load is of primary importance. A doubling of the load over the same period of time results in

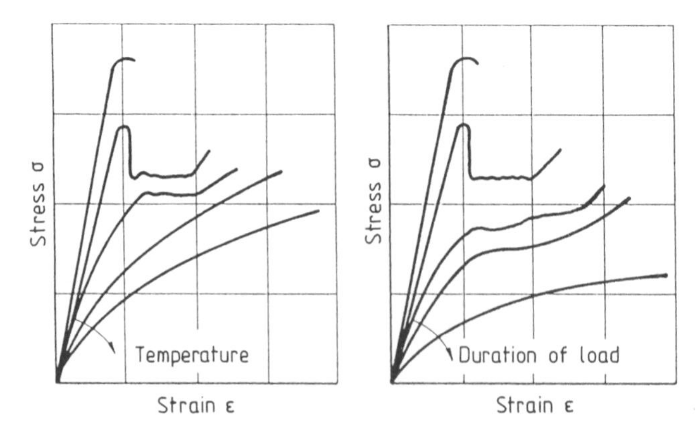

Figure 4.1.1 Effect of temperature and time on the deformation behavior.

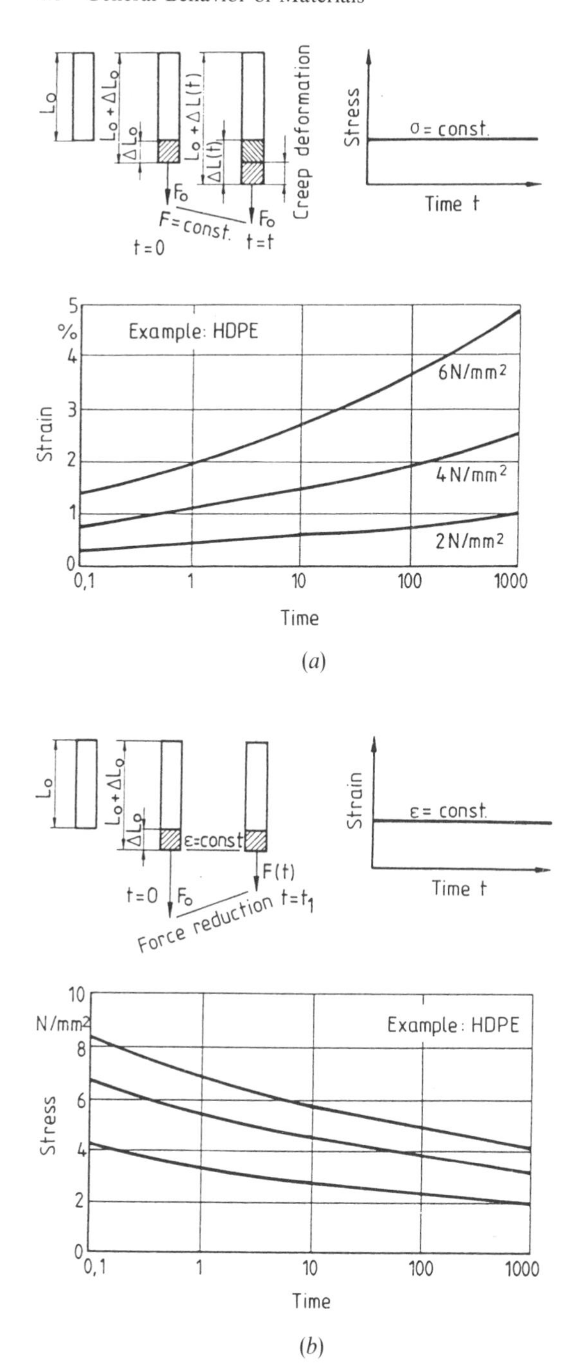

Figure 4.1.2 (a) Viscoelastic material behavior (retardation). (b) Viscoelastic material behavior (relaxation).

more than twice as much *deformation*, and a doubling of the deformation over the same period of time results in less than twice the stress (isochronic stress–strain diagrams are thus degressive).

- Plastics exhibit temperature-dependent materials behavior.

At an elevated temperature, the time-dependent phenomena are accelerated. This accelerating effect is described empirically by *time–temperature shift principles*, e.g., according to Arrhenius.

This behavior, especially the time-dependence of many material properties for thermoplastics in particular, makes it necessary to find *accelerated test methods* and reliable *extrapolation methods*.

To be able to extrapolate to the long-term behavior under practical conditions, "*aggravated*" *tests* are conducted at higher stress levels, temperatures, or chemical concentrations. A prerequisite for extrapolation is the absence of any significant material degradation.

As an accelerating method for the *aging behavior*, i.e., especially with regard to the degradation of materials upon exposure to oxygen and heat, extrapolation according to the so-called *Arrhenius Law* has proven successful. In this case, the accelerated test takes place at temperatures considerably above the operating temperature—but below the transformation range—and the results are extrapolated to longer lasting loads at lower temperatures.

Test specimens are maintained at different temperatures, and the time required for a certain property to decrease by a specified percentage is determined. The induction period required for the specified property change is plotted logarithmically versus the reciprocal temperature.

The property changes measured, e.g., the elongation at fracture, fall on a straight line in such plots. From these the life expectancy at the operating temperature can be extrapolated.

One method for describing the material behavior is the so-called *deformation model*, which is based on a multitude of parallel *non-linear Maxwell elements* (Fig. 4.1.3). Each of these elements is described by the spring stiffness and the flow function associated with the dashpot. This provides the relationship between the stress acting on the dashpot and the resulting flow rate.

On the whole, plastics are quite correctly called "tailor-made materials." When selecting materials, this special property profile must be matched to the often complex set of

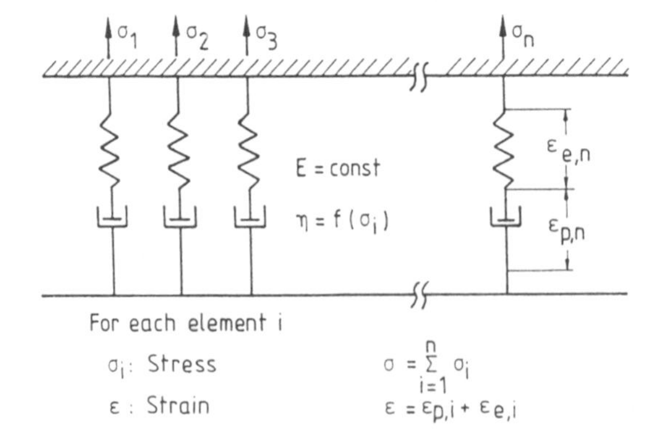

Figure 4.1.3 Deformation model [according to Menges (1990) and Schmachtenberg (1985)].

requirements for the application. The necessary material characteristics are determined in materials tests based on standardized methods and are thus comparable.

4.2 Stress–Strain Behavior

In almost all applications of plastics, the mechanical properties in some form are of major importance. With regard to the mechanical properties, especially the stress–strain behavior, the magnitude and duration of the applied load play a major role, along with the external factors, such as temperature and surrounding substances, acting on the material. This is considerably more pronounced with the *thermoplastics* than with the *thermosets* or the *fiber-reinforced plastics*. Accordingly, the usual specification of strength values, such as the *tensile strength*, or of elastic characteristics, such as the *modulus of elasticity*, is not adequate for evaluation of the mechanical behavior of plastics materials. It is only the stress–strain behavior that describes the relationship between internal material stress resulting from externally applied forces and the induced material strain. This represents important information for the design engineer.

With regard to types of loads, a distinction is made between *unidirectional* and *bidirectional loads*, with the unidirectional load preferred for testing purposes and bidirectional loads applied only in special cases. In general, test specimens are investigated in bending, compression, tensile, and shear tests with respect to their behavior under these types of loads. These common test methods are described in various standards and test procedures.

The response of the test specimen to a uniaxial load condition is described most simply with the aid of the tensile test, for which reason the following is based primarily on this type of load. With *bending, compression, and shear loads*, a superposition of stresses in the material, especially with *anisotropic* materials such as laminated or filled systems, can be expected in spite of the quasi-unidirectional application of force. In a tensile test, a test specimen is stressed at a specified strain rate by an externally applied force acting parallel to the longitudinal axis of the specimen, with no external forces perpendicular to the direction of load and thus nothing to hinder elongation. During the load cycle, elongation is measured and plotted as a function of the tensile force. In this way, a *force–deformation curve* is obtained, which can be converted into a so-called *stress–strain diagram* either by calculation or through suitable measurement techniques that take into consideration the important test parameters.

4.2.1 Short-Term Behavior

When *dimensioning* plastics products, values from short-time tests that were determined under the types of loads described above serve as the basis, just as with other materials. The basic forms of stress–strain diagrams for brittle, tough/elastic, and soft/elastic materials are shown in Fig. 4.2.1. In short-time tests, the *rate of load application* is so selected that a rapidly increasing stress or deformation occurs in such a manner that the characteristic material property—as a rule, *destructive* or *unstable specimen failure*—is established within a few minutes. The loading rate can be achieved through a constant increase in load or through a constant *deformation rate*, with the latter being common in the testing of plastics.

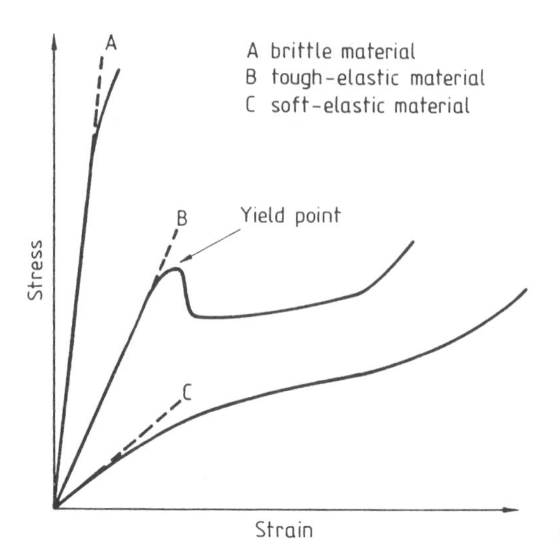

Figure 4.2.1 Basic shapes of stress–strain curves.

Since plastics belong to the viscoelastic materials, their mechanical properties depend on the duration and rate of loading. Fig. 4.2.2 shows the effect that different strain rates have on the tensile strength of several thermoplastic resins. It can be seen clearly that the various plastics exhibit a maximum strength at characteristic strain rates. These strain rates sometimes lie considerably above those selected for *short-term loading*.

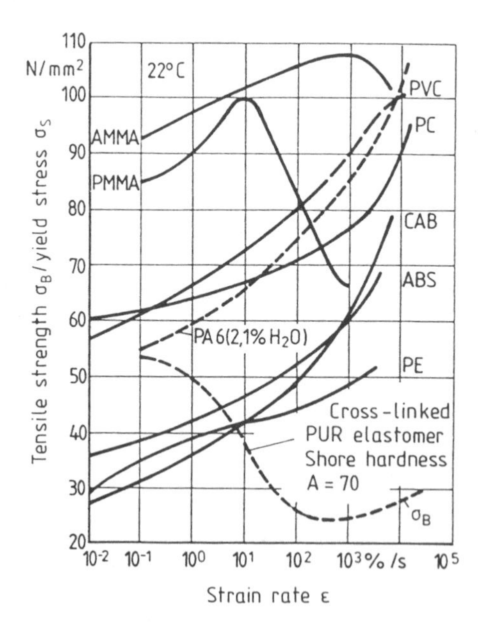

Figure 4.2.2 Dependence of tensile strength and yield stress on the strain rate [according to Oberbach (1975)].

4.2.2 Impact Loading

When the load is applied so rapidly as to be instantaneous, the material can react in such a brittle manner that deformation-free or low-deformation *glasslike fractures* occur. An impact-resistant molded part is expected to absorb energies that occur under impact loads elastically or plastically without fracture, depending on the field of application. Materials that withstand the greatest possible deformation at high forces have high *energy absorption*. In other words, the area under the stress–strain curve (integral of the stress over the strain) corresponds to the work expended. In this way it is possible to obtain initial information on the *toughness properties* from the results of tensile tests at various strain rates. The relationships between stress, strain, strain rate, and testing temperature are shown in Fig. 4.2.3 for thermoplastics.

To investigate the toughness properties of plastics under instantaneous (shocklike) loads, the standardized *impact resistance* and *tensile impact tests* are employed for the most part. For investigations under two-dimensional loads, the standardized *penetration test (falling dart test)* has also proven suitable. For test specimens that do not break under impact or tensile impact testing, the *impact resistance* cannot be expressed numerically; they are nevertheless to be classified as impact-resistant. Such materials are given a defined notch and subjected to so-called *notched impact strength tests*. Notches can be U-shaped, V-shaped, or circular.

Impact tests can be conducted with appropriately equipped testing instruments in a manner similar to that for short-time tests and common types of loads. Such tests have demanding requirements with regard to the resolution of the readings with time. As a result, these tests have generally not yet been standardized and are more scientific in character.

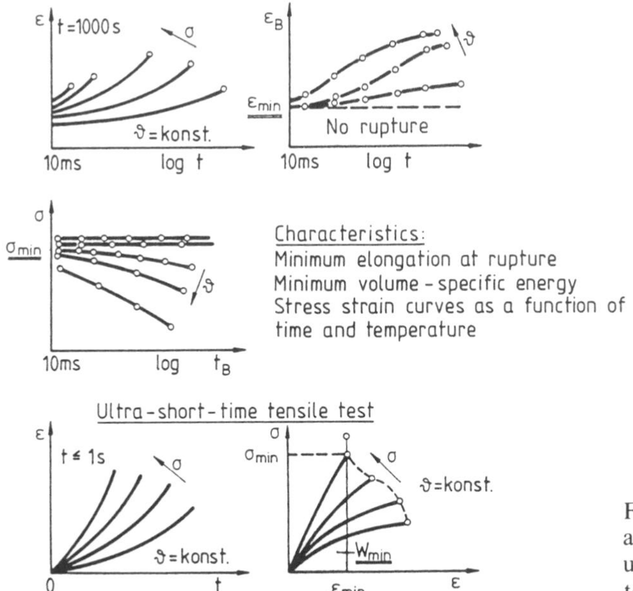

Figure 4.2.3 Deformation and failure of thermoplastics under impact loads [according to Menges (1990)].

4.2.3 Behavior under Long-Term and Static Loads

For the engineering application of plastics under long-term loads, the events during so-called *creep tests* are important. In the following, the effects of the duration and magnitude are discussed specifically. The effect of aging and effects due to surrounding substances that differ from those encountered under normal conditions are not discussed here. These must be given special attention in actual applications, since they can be quite considerable.

To describe the long-term behavior under a static load, the following diagrams are generally employed:

- time–deformation curves (time-elongation curves),
- creep modulus curves,
- time–stress curves,
- isochronic stress–strain diagrams.

The *time–deformation curves* or *time–elongation curves* are determined in long-term bending, compression, tensile and shear tests, in other words, the so-called *retardation tests*. They show the deformation behavior of materials under specific external conditions. As a rule, data are presented in logarithmic or log–log plots, with the logarithmic time axis along the abscissa and the deformation variable along the ordinate. Various stress states and other ambient conditions such as temperature or surrounding substance can appear as parameters. The values from the retardation tests are obtained by subjecting the material or test specimen to a constant load from a given height and recording the time-dependent deformations. Fig. 4.2.4 contains illustrations of several creep curves. Since plastics products are supposed to have a long period of utility under certain circumstances, and creep tests naturally require long periods of time, extrapolation is permissible if the creep curves are linear (in a

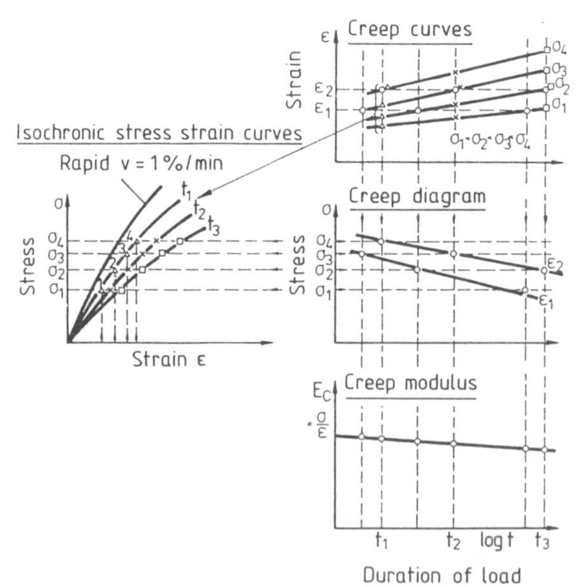

Figure 4.2.4 Preparation of creep diagrams.

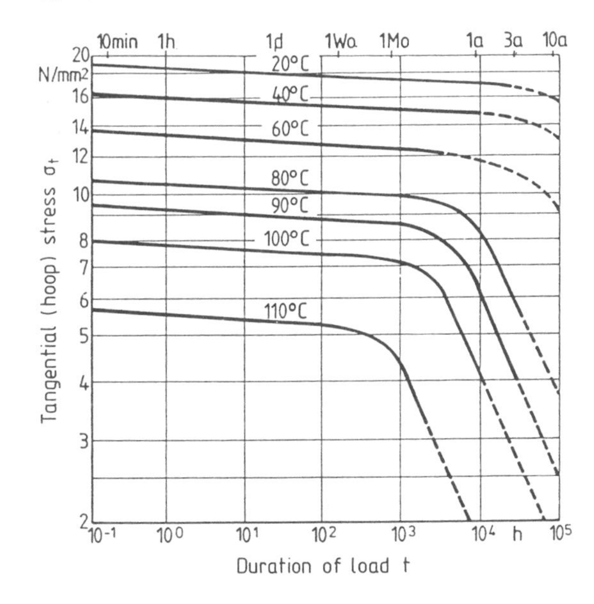

Figure 4.2.5 Creep strength of polybutylene pipes under internal pressure. Test medium: water.

semi-logarithmic plot) and the critical elongation limit (limit of elasticity) is taken into consideration.

The *creep modulus curves* describe the time dependence of the "elastic" properties of plastics. They are obtained by converting the results of time–deformation curves or time–stress curves. The term "creep modulus" is not directly comparable to the term "elastic modulus"; it is merely a correlation factor between a continuously applied stress and the resulting time-dependent strain.

The *time–stress curves* are obtained either from *relaxation tests*—here, the decrease in stress is determined as a function of time at a specified constant deformation—or by simply replotting several time–deformation curves. Data are presented in a semi-logarithmic or log–log plot analogous to those for the time–deformation curves. The family of curves shows how the stresses decrease with time at a constant deformation. The time to rupture increases with decreasing stress. For plastics simultaneously subjected to a static load and the effects of temperature and surrounding substances, determination of *creep strength curves* from tubular-shaped test specimens has proven useful (Fig. 4.2.5). Knowledge of the *creep strength*, especially of elastomeric materials, is not adequate for dimensioning components subjected to long-term loads. In general, the lower the strength of a product, the greater is its elongation at rupture. For low-strength plastics, the elongational behavior thus must also be taken into consideration.

Isochronic stress–strain diagrams have proven useful sources of information for designers. They are obtained by exchanging parameters in the time–elongation and time–stress curves.

The entire functional relationship between stress, deformation, and time can be represented clearly in a *creep diagram*. Permissible stresses and deformations as well as deformations at given stresses or stresses at given deformations can be obtained while considering the effect of time and, possibly, temperature. The relationships and types of graphic conversions can be seen in Fig. 4.2.4 for constant temperature conditions.

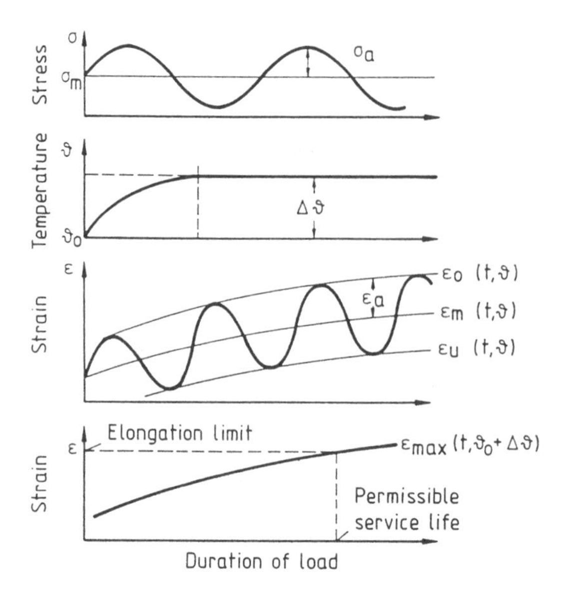

Figure 4.2.6 Creep under a dynamic load [according to Thebing (1979)].

4.2.4 Dynamic Loads

When periodically changing loads occur, which is often the case in machinery construction and automotive engineering, the strength values that were determined under a constantly increasing load and deformation or in a creep test cannot be used for calculating the individual design components. Repeated loading can lead to *component failure* at lower stresses and deformations than is the case for static loads.

The behavior of materials under oscillating (cyclic) loads is determined in *flexural fatigue tests*, with the parts to be investigated usually subjected to a sinusoidally applied load. The viscoelastic component in the behavior of plastics leads to an irreversible absorption of energy under load that can lead to heating of the material in a very brief period of time under repeated stress. This phenomenon, which is affected in part by the testing frequency and quite decisively by the conditions under which the generated heat is removed, must be taken into consideration when performing and evaluating flex tests. With *flexural loads*, a distinction is made between *alternating stresses* and *stress limits*, which can have different effects on the fatigue behavior. In addition, a distinction is made between constant stress amplitude and constant strain amplitude. In Fig. 4.2.6 the effects of constant sinusoidal alternating stresses, and of the resultant heat generated, on the creep, i.e., on the elongation behavior, are illustrated by way of example. In plastics as well, the *Wöhler curves* known from metals asymptotically approach a stress limit with increasing number of cycles; this limit is generally called the *fatigue limit*.

4.3 Hardness, Wear, and Frictional Behavior

In the following three subsections, the question of the behavior of plastic surfaces under contacting mechanical loads is discussed, along with its importance for practical use.

4.3.1 Hardness

The term *hardness* generally describes the resistance of one body to penetration by another body. Often hardness is considered in conjunction with *wear and frictional behavior*. There is no standard unit of measure for this property, i.e., the hardness is a characteristic value.

In principle, the test methods for hardness (see Fig. 4.3.1) can be classified into three groups:

- methods that measure the overall deformation, e.g., *spherical indentation* and *Shore hardness*;
- methods that measure the plastic deformation, e.g., *Rockwell* and *Vickers hardness*;
- methods that determine the elastic response, e.g., *impact elasticity, resilience*.

In addition to the temperature and duration of load, it is necessary to specify exactly the geometry of the indenter as well as the type and magnitude of load.

Fig. 4.3.2 presents a comparison of the indentation methods, along with a true scale representation of indenter shape, range of indentation depth, and directly observed surface region. The values obtained from the various methods can hardly be compared with one another, but they nevertheless often provide useful information for a materials comparison or materials selection. Hardness values are not design values, however, i.e., they cannot be

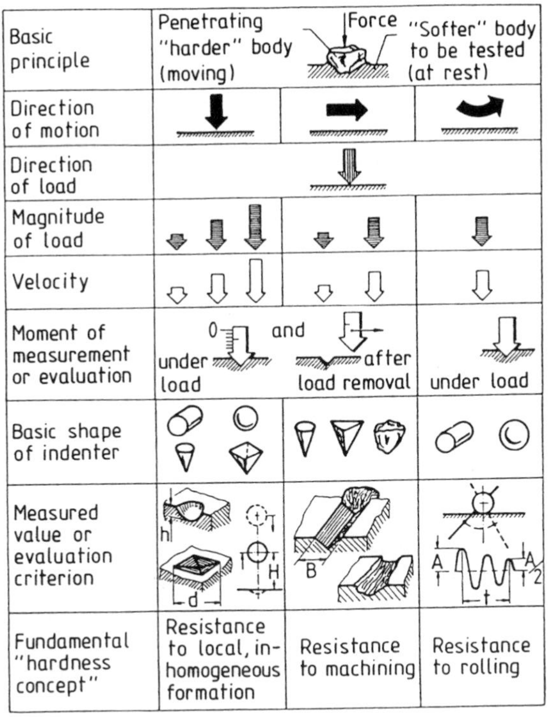

Basic principle	Penetrating "harder" body (moving)	Force	"Softer" body to be tested (at rest)
Direction of motion			
Direction of load			
Magnitude of load			
Velocity			
Moment of measurement or evaluation	0 under load	and after load removal	under load
Basic shape of indenter			
Measured value or evaluation criterion	h H d	B	A t A/2
Fundamental "hardness concept"	Resistance to local, inhomogeneous formation	Resistance to machining	Resistance to rolling

Figure 4.3.1 Principle methods of "hardness" testing [according to Müller (1985)].

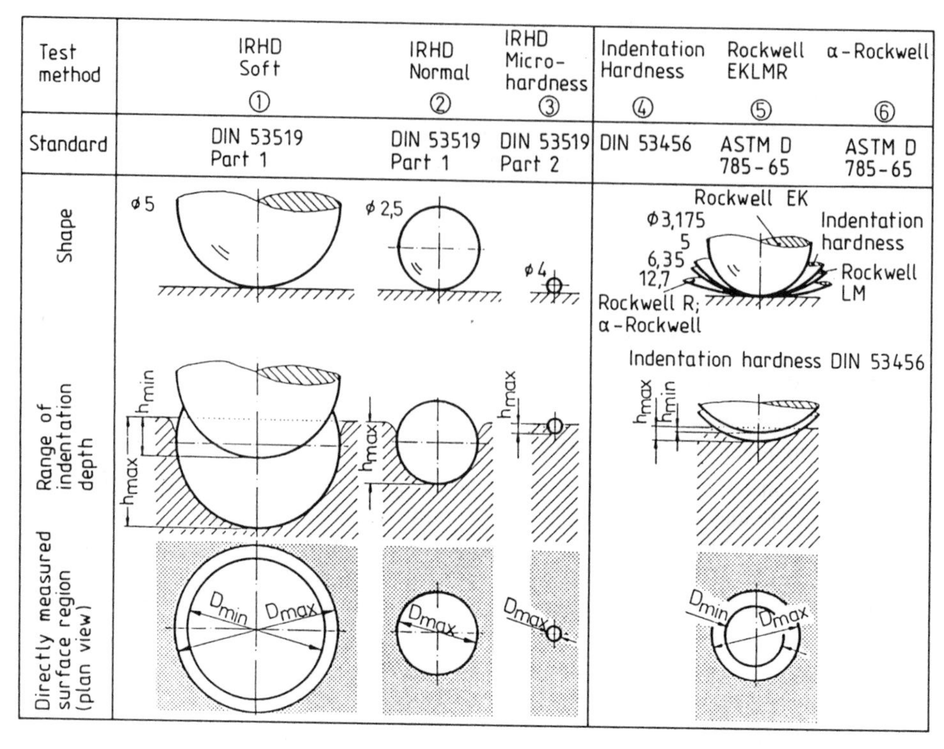

Test method	Shore A ⑦	Shore D ⑧	Barcol ⑨	Vickers ⑩	Knoop ⑪	Buchholz ⑫
Standard	DIN 53 505	DIN 53 505	EN 59	DIN 51 225	ASTM D 1474	DIN 53 153
Shape						
Range of indentation depth	h_{min}, h_{max}, x, y	h_{min}, h_{max}, xy	h_{max}	h	h	h_{max}
Directly measured surface region (plan view)	x, y	x, y				

Figure 4.3.2 Hardness test methods for plastics 4–6, 8–11; rubber 1–3, 7 and coatings 11, 12 [according to Müller (1985)].

taken as the basis for design calculations. They have proven useful for production control and incoming materials inspection. Developments in the field of *microhardness* measurements in particular hold the promise of nondestructive quality monitoring and of being a means of assessing aging as a result of weathering, thermal action, or the action of surrounding substances.

4.3.2 Wear

The *scratch resistance test* is a precursor to wear testing. *Attrition*, or *wear*, is defined in DIN 50 320 as the undesirable surface change resulting from dislodgement of particles as the result of mechanical causes. The difficulty with regard to testing and evaluation lies in the great number of possible combinations of primary factors (Fig. 4.3.3). Various test methods have been developed taking into account the special load conditions.

In technical applications of plastics, the question of attrition, or wear, is an important aspect, e.g., for coatings, bearings, or gears. The test methods listed can provide comparative values. Attention must be given to the surface finish, temperatures, velocities, and effects of surrounding substances. For a comparison, it is important that all major wear factors be in agreement.

4.3.3 Frictional Behavior

The term *friction* describes the resistance to relative motion of two bodies in contact with one another. The above-mentioned attrition, or wear, occurs when the tangential or normal forces acting during the relative motion destroy both or one of the materials involved.

The inclined plane method is the simplest for determining the *static* and *kinetic coefficients of friction*. Special test methods have been developed for film, sheet, and molded parts, and are employed for comparative materials evaluation. The *surface finish* of the sliding materials, *contact pressure*, velocity, temperature, lubrication (interfacial substances), and miscellaneous surrounding influences are important factors that must be observed. The *stick-slip effect* that is observed with some plastics and under certain conditions of use should be mentioned for *dry sliding conditions*.

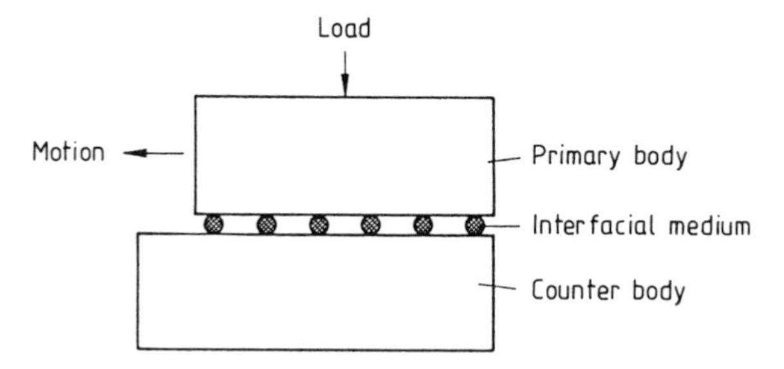

Figure 4.3.3 Model for analysis of the wear process [according to Racké].

4.4 Electrical Behavior

4.4.1 Dielectric Behavior

The electrical properties of plastics are determined primarily by the mobility of the molecular constitutional units. The relative *dielectric constant* ρ is the characteristic quantity for this property. It indicates by what factor the capacitance of a capacitor changes, with respect to the capacitance in vacuum, if the plastics material is placed between the plates of the capacitor:

$$C = \varepsilon_r \cdot C_0 = \varepsilon_0 \cdot \varepsilon_r \frac{A}{d}$$

with

C_0	Capacitance in vacuum
ε_0	Dielectric constant for vacuum
A	Area of the capacitor
d	Distance between the plates

where C_0 is the capacitance in vacuum, ε_0 is the dielectric constant for vacuum, A is the area of the capacitor, and d is the distance between the plates. The change in capacitance of a capacitor as the result of inserting a plastics resin (*dielectric medium*) is caused by polarization charges that form in the electric field of the capacitor. Depending on the structure of the plastics, the polarization can occur via various mechanisms.

Electron polarization: Completely nonpolar plastics, e.g., polyethylene, polyisobutylene, and polytetrafluoroethylene, can exhibit only electron polarization. The electron shell of the atoms is displaced by the electric field. For the case of electron polarization alone, the dielectric constant ε_r is related to the optical index of refraction according to the following equation:

$$\varepsilon_r = n^2$$

In the case of electron polarization alone, the dielectric constant is practically independent of the frequency and decreases slightly with temperature, since the number of polarizable particles per unit of volume decreases because of thermal expansion.

Ion polarization: Ion polarization hardly ever occurs in polymers.

Dipole polarization: In polar plastics, the entire molecules cannot become aligned in the electrical field. Polarization results from an *orientation* of chain segments or side chains. Dipoles can also become oriented independently of the main chain if rotation around the chain is possible. The dielectric constant is larger for polar plastics than for nonpolar plastics.

In an alternating electric field, the motion of the molecular segments leads to internal friction and thus heating of the plastic. For the case of electron polarization alone, this internal friction is very slight. The internal friction is identified by the *dissipation factor* $\tan \delta$. The dissipation factor describes the phase shift between the current and the voltage. In a loss-free capacitor, the dissipation factor $\tan \delta = 0$, i.e., the current leads the voltage by 90°. The

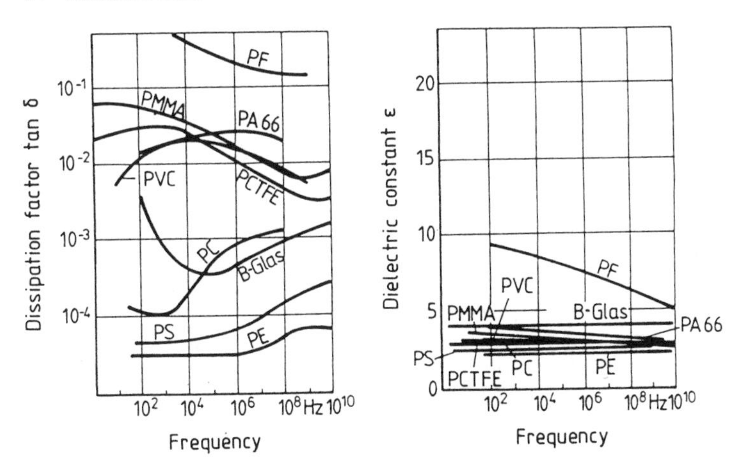

Figure 4.4.1 Frequency dependence of the dissipation factor tan δ and of the dielectric constant ε of plastics at 20°C [according to Domininghaus (1986)].

dissipation factor is frequency-dependent. In the *dispersion region*, the frequency dependence is particularly pronounced (Fig. 4.4.1). Semi-crystalline plastics, which exhibit an amorphous and a crystalline phase, have two dispersion regions that are characteristic for the two phases below the crystallite melting point. Plasticizers result in the decoupling of neighboring chains. This results in an increase in the *dispersion frequency* at a given temperature. Cross-linking results in a stiffening of the molecular structure so that the loss maxima are shifted to higher temperatures for a given frequency. Depending on the field of application, attention must be given to whether the dissipation factor is low or high.

In *insulators* for high-frequency currents and radar antenna coverings, the lowest possible dissipation factor is desired in order to have only low power losses. Polyethylene and polystyrene have proven especially useful in these applications. On the other hand, *high-frequency welding* is possible only if sufficient heating of the plastic materials is assured. For insulators, the dissipation factor tan δ should be $< 10^{-3}$, while for high-frequency welding tan δ should be $< 10^2$.

4.4.2 Electrical Conductivity

Plastics are insulators with very high *resistance*. A distinction is made between the *volumetric resistance* and the *surface resistance*, which are measured in accordance with the German standard DIN 43 482. The resistance values are measured 1 min after applying the voltage, since the d.c. resistances are time-dependent. The mechanism of conductivity is very complicated, since *interfaces* between the individual phases (amorphous and crystalline), fillers, plasticizers, moisture, and emulsifiers contribute to the conductivity. It can be assumed that *ion conduction* contributes significantly to the conductivity. Ion transport can be attributed to *diffusion*, which is possible in all plastics. Two types of transport are possible:

- For migration via displacement, there must be *free volume* present and a *potential barrier* must be overcome. Free volume is increased by raising the temperature; the potential barrier can be overcome by applying an electric field. The electrical resistance of plastics thus decreases with increasing temperature.

- For transport along interfaces, which plays a role especially for surface resistance, *potential thresholds* must also be overcome, which is possible through thermal motion. The absorption of water is of major importance. This leads to a reduction of the potential thresholds. The surface resistance is high for *hydrophobic* substances, and low for *hydrophilic* ones. *Electron conduction* plays no role in plastics, since there are no electrons in the conduction band. The potential threshold cannot be overcome through thermal energy. The conductivity can be increased significantly through irradiation with high-energy beams. For certain purposes, e.g., dispersive floor coverings, *conductive plastics* are desirable. This conductivity can be achieved through incorporation of *conductive fillers*, such as special carbon blacks, for instance.

4.4.3 Dielectric Strength

The dielectric strength plays an especially important role for insulation against high voltage. The dielectric strength is measured in accordance with DIN 53 481 using plates with geometrically different electrode arrangements. In the presence of an applied electric field, the conductivity results in a power conversion that is proportional to the square of the field strength. The conductivity and power generated increase exponentially with the absolute temperature. In case of insufficient heat removal, plastics will be destroyed. The channel formed exhibits high conductivity even when cold, since the track consists primarily of carbon from the decomposition products of the plastic. The voltage that can be withstood increases with the square root of the plate thickness. This means that higher field strengths can be withstood by thinner films (Fig. 4.4.2).

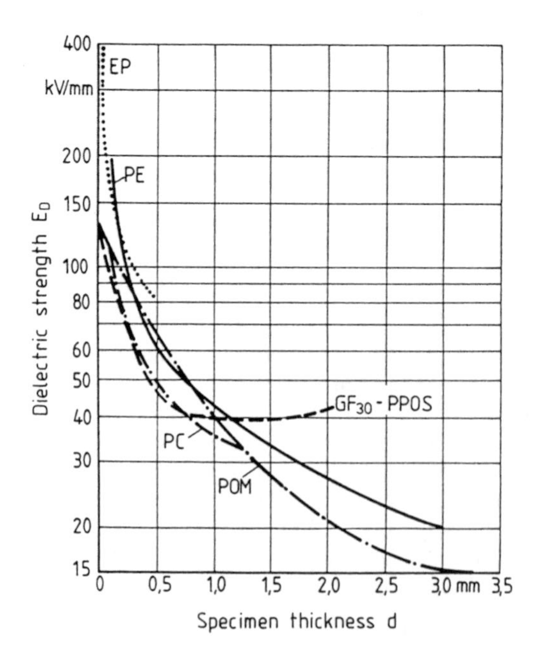

Figure 4.4.2 Dielectric strength as a function of specimen thickness [according to Oberbach (1975)].

4.4.4 Electrostatic Charges

Electrostatic charges on plastics are often an undesirable side effect. The risk of sparking, malfunctioning of measurement equipment, and dust deposits are possible consequences. Electrostatic charges resulting from friction with other solids are retained for a long time by poor conductors such as plastics and are discharged only through contact with conductors or oppositely charged materials. Depending on the geometry of the surfaces and the size of the gap, sparking may already occur at a field strength of 5 kV/mm.

In order to avoid this sparking, the *volumetric resistivity* must be reduced to at least $10^9\ \Omega$ cm. Another possibility for reducing dangerous electrostatic charges is to reduce the *surface resistivity*. This is possible by creating a hygroscopic surface, e.g., with a soap solution. Alternately, the surface resistivity can be modified by incorporating incompatible substances that migrate to the surface.

4.5 Optical Behavior

4.5.1 Refraction and Dispersion

The optical properties of a material involve its behavior with respect to *light rays*, i.e., with respect to rays in the visible portion of the spectrum as well as in the ultraviolet and infrared regions. In passing from one material to another (with a different index of refraction), the light waves undergo a change in direction, or *refraction*. Snell's Law applies in isotropic substances:

$$\frac{\sin \alpha}{\sin \varepsilon} = n$$

where α and ε are the angles with respect to the vertical formed by a light beam in the first and second materials.

As a physical quantity, the *index of refraction n* is important in distinguishing plastics, especially in the application of organic glasses for optical purposes. It must be borne in mind that the indices of refraction for all materials depend on the frequency of the light. This effect is called *optical dispersion*.

4.5.2 Transparency

Compared to metals and wood products, many plastics exhibit the advantage of *transparency*. The optical properties of amorphous thermoplastics such as PC, PMMA, PVC, and UP resins do not differ significantly from those of inorganic glasses (Fig. 4.5.1). Transparency, *turbidity*, and *clarity* are interrelated terms. If an image generated by an optical system exhibits distortions, or points that were distinct in the original run together in the image, there is said to be a decrease in clarity. *Optical scattering* with a severe loss of contrast results in turbidity. The ratio of the light transmitted without reflection to the intensity of

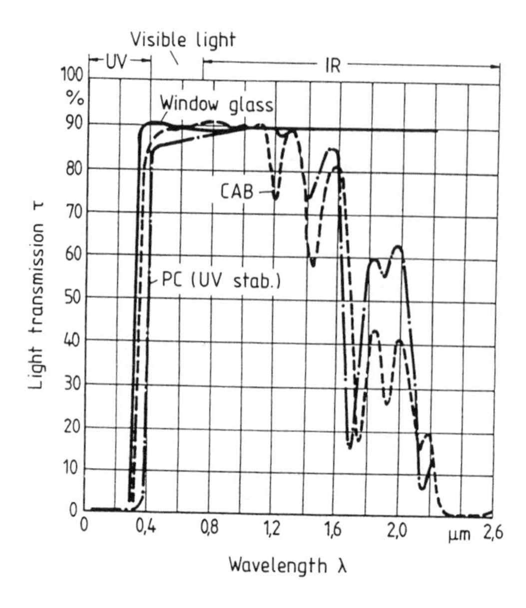

Figure 4.5.1 Dependence of light transmittance on the wavelength [according to Oberbach (1975)].

the incident light is called the transparency. Turbidity and reduced clarity indicate irregularities in the surface. These may involve filler particles, dirt particles, defects, surface roughness, and the like. Environmental factors—weathering, thermal cycling, etc.—can cause a change in turbidity and clarity. Examples of this include double-skin panels and skylights of PVC and PC, which experience a loss in clarity with increased weathering.

4.5.3 Gloss

Gloss is not a purely physical quantity, but rather is also physiological and psychological in nature. It is thus not possible to measure the gloss directly, but it is possible to determine as a *reflectometer value* the "gloss capacity," namely the contribution of the surface, as a result of its *reflection properties*, to the creation of the impression of gloss. The gloss of plastics articles can be influenced during production (e.g., by the mold/die surface), through secondary treatment, via the *surface structure*, or through appropriate additives.

Glossy surfaces are generally less likely to become soiled than matte ones. A variety of factors such as weathering, temperatures, effects of surrounding substances, and mechanical loads lead to a reduction in the gloss of automotive finishes, paint films, and almost all plastics articles encountered in daily use.

4.5.4 Color

Depending on the type of plastic and the desired optical effect, soluble colorants, so-called *dyes*, are employed to impart color to plastics. Dyes impart color to the plastic solely as the result of *absorption* by the dye molecules. They are preferred for fully transparent colors. For semi-transparent to opaque colors, colorants are employed that contribute to the

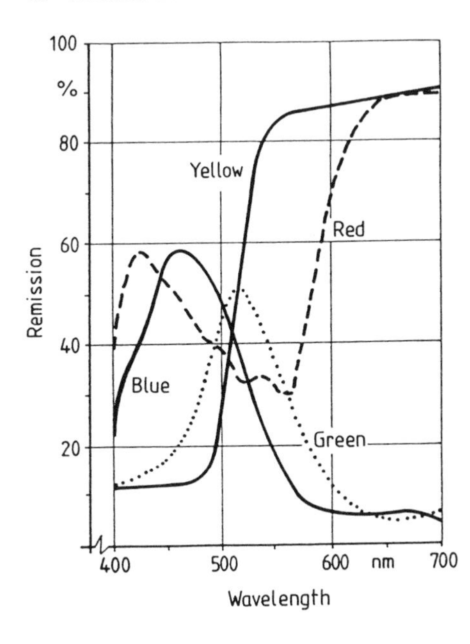

Figure 4.5.2 Remission values of specimens with different colors.

appearance of the color through absorption and scattering by the colorant particles. Color is an impression conveyed by the human eye and is the result of partial reflection, partial transmission, and partial absorption of the light emitted from a light source. The perception of color thus depends on three factors:

- the type of illumination, i.e., the relative spectral distribution of the illuminating light source;
- the optical properties (*remission* and *transmission values*) of the illuminated object;
- the spectral sensitivity of the human eye.

The starting point for measuring the color of a sample is recording of the remission or transmission curve, i.e., the intensity of the reflected light as a function of the wavelength in conjunction with the illuminating light source (Fig. 4.5.2). The *standard color values* X, Y, Z, and the resultant *standard color value fractions* x, y, (z), are obtained by mathematically relating such curves to the so-called *standard spectral value curves*. The x and y values can be plotted in a rectangular coordinate system (*CIE color table*, CIE = Comission Internationale de l'Eclairage). The region of attainable colors is defined in this representation by a surface resembling the shape of a "shoe sole" in the x, y diagram. In this CIE color system, colors can be represented and compared with one another.

4.6 Acoustic Behavior

The *acoustic behavior* of plastics is closely related to the *dynamic-elastic behavior*. There are, however, characteristic differences between solid and foamed plastics. The *modulus of elasticity* E and *Poisson's ratio* are the primary quantities with which the acoustic behavior can be described. The other common moduli depend on these two quantities. The modulus of elasticity E determines the propagation of dilatational (extensional) waves in a rod, while the *longitudinal wave modulus* L describes the propagation of compressional waves in an endless medium. The *shear modulus* G describes the propagation of transverse, torsional, and shear waves in rods or in endless media in which the direction of oscillation is perpendicular to the direction of propagation. The modulus of elasticity and Poisson's ratio can be used to express these moduli according to the following equation:

$$G = \frac{1}{2} \cdot \frac{E}{1 + \mu}$$

$$L = E \frac{1 - \mu}{(1 + \mu)(1 - 2\mu)}$$

Each of these moduli is associated with the corresponding *speed of sound*:

$$c_D = \sqrt{\frac{E}{\rho}} \qquad c_T = \sqrt{\frac{G}{\rho}} \qquad c_L = \sqrt{\frac{L}{\rho}}$$

c_D speed of sound for a dilatational wave,

c_T speed of sound for a transverse wave,

c_L speed of sound for a longitudinal wave.

Poisson's ratio μ is about 0.3 for plastics in the glassy state and about 0.5 in the elastomeric state. This results in different speeds of sound (Fig. 4.6.1). When considering the propagation of *sound waves*, the *attenuation* resulting from internal friction must be taken into account. The mechanical quantity used to describe the attenuation is the *loss factor*. The loss factor assumes especially high values for regions involving changes of state (e.g., solidification

	Modulus [N/mm²]	Wave velocity [m/s]
Glassy state, $\mu = 0.3$		
Modulus of elasticity	$E \approx 10^3$ to 10^4	$C_D \approx 200$
Shear modulus	$G = \frac{E}{2(1+\mu)} \approx 3.8 \cdot (10^2 \text{ to } 10^3)$	$C_T \approx 1000$
Longitudinal modulus	$L = \frac{(1-\mu) \cdot E}{(1+\mu)(1-2\mu)} \approx 10^3$ to 10^4	$C_L \approx 2000$
Rubberlike state, $\mu = 0.5$		
Modulus of elasticity	$E \approx 1$ to 10^2	$C_D \approx 10$ to 400
Shear modulus	$G \approx \frac{E}{2 \cdot (1 + 0.5)} \approx \frac{E}{3}$	$C_T \approx 6$ to 200
Longitudinal modulus	$L \gg E$	$C_L \approx 2000$

Figure 4.6.1 Moduli and speed of sound as functions of the physical state [according to Menges (1990)].

region), and with a value of 10^{-2} already in the glassy condition is large compared to metals with 10^{-5} to 10^{-3} or glass with 10^{-3}. This leads to sound waves with rapidly decreasing amplitudes in plastics.

The *reflection* R of a sound wave is described by the following equation:

$$R = \frac{Z_2 - Z_1}{Z_2 + Z_1}$$

In the above, $Z = \rho\, c$, the *impedance*, with c the speed of sound for the type of wave considered (see above). The index 2 is associated with the reflecting substance, the index 1 with the substance from which the sound wave impinges against the reflecting wall. When the incident sound is transmitted via a gas, plastics behave like metals; the sound is largely reflected. When the incident sound is transmitted by a liquid, however, Z_1 is approximately equal to Z_2, resulting in little reflection. Metals also reflect waves from liquids almost completely (Fig. 4.6.2).

In technical acoustics, a distinction must be made between *sound attenuation*, or *reflection*, and *sound damping*, or *propagation*, and *absorption*. A high mass, i.e., a heavy wall, is required for *sound reflection*. If the objective is to avoid *sound propagation* in a solid body, soft intermediate layers are required. Elastomeric supports for machinery are an example of this.

So-called *sound-absorbing materials* are open-cell foams that cannot be considered rigid. The wave enters the cells and loses its energy through friction between the molecules in the air and through deformation of the walls of the sound-absorbing material. In the case of *sound-damping substances*, plastics are used that have a high loss factor over the range of service temperatures and that adhere well to the material in which the sound is to be damped.

For *ultrasonic welding* it is necessary to direct the energy of the sound waves to the joint surface without significant losses and convert it there into heat. This means that materials with a high modulus of elasticity and small loss factor can be welded more readily with this method. Conversion of the stimulated vibrations into heat is brought about in part through *frictional losses* at the joint surface. An additional significant amount is provided by internal damping in the material, with this effect being concentrated in the immediate vicinity of the joint zone through appropriate shaping of the near-joint regions of the parts to be joined.

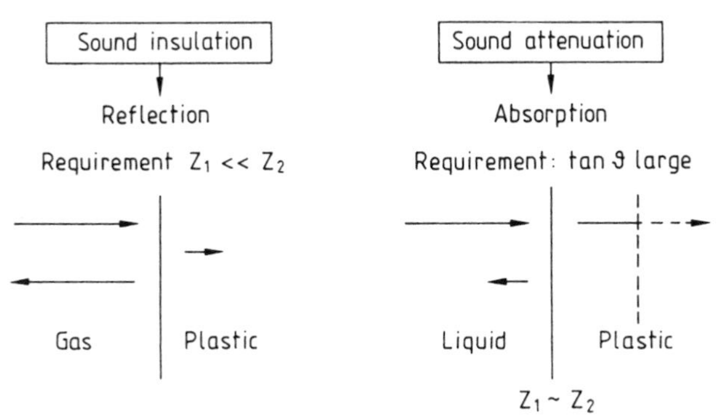

Figure 4.6.2 Sound insulation and sound attenuation in plastics.

The molten zones absorb the ultrasonic energy particularly well because of the increased loss factor, which generally rises with temperature. It can be noted in general that amorphous thermoplastics can be more easily welded than semi-crystalline ones.

4.7 Response to Environmental Factors

4.7.1 Resistance to Chemicals

When considering the action of chemicals on plastics, the *diffusion* (of components) of the chemical into the material plays a much greater role than is the case for corrosion of metals, where the processes remain limited largely in the near-surface regions. For this reason, a liquid substance can give rise to a change in the properties of the plastic even without a chemical reaction.

The *resistance* to chemical attack is determined primarily by the chemical composition of the plastic. Polymeric materials, the macromolecules of which consist of hydrocarbon chains (PE, PIB), are very resistant to acids, bases, and weak oxidizing agents. With the introduction of substituents into the polyethylene chain—hydroxyl, acetyl, or other functional groups —the *chemical resistance* is lowered. However, if, on the other hand, the hydrogen in the polyethylene is replaced with fluorine, the resultant polytetrafluoroethylene (PTFE) is resistant to practically all chemical *agents* and is to date unmatched by any other plastic in this respect. In principle, every chemical attack can cause *irreversible changes* to the plastics molecules, such as chain degradation, *cross-linking*, changes in the chemical composition of the molecular chains, e.g., oxidation, etc.

The action of physically active substances is based on an expansion of the molecular network (visible externally as *swelling*). As a result, the mobility of the macromolecules increases, the hardness and strength decrease, and the electrical and other physical properties change. The degree of change depends on the *polarity* of the material and chemical substance together (like dissolves like):

- Nonpolar polymers such as PS, PE, and PIB swell or dissolve in nonpolar *solvents* (e.g., gasoline, benzene), while they are resistant to polar solvents (water, alcohol).
- Polymers that contain polar groups, such as polyvinyl alcohol, polyamide, and the like, are resistant to nonpolar substances, swell, and dissolve in polar solvents.

The resistance to chemical substances is determined by suspending specimens in these materials without any mechanical load. After increasing exposure times, specific property changes are measured and the results are evaluated according to a general ranking such as resistant, conditionally resistant, and not resistant, for instance (Fig. 4.7.1). Such a classification naturally gives only a rough indication of the suitability of a plastic for a specific application. Testing under actual conditions is indispensable in many cases.

4.7.2 Environmental Stress Cracking

In metals the phenomenon of *stress corrosion cracking* is well known. It leads to *deformationless inter-* or *transgranular* separation in the simultaneous presence of certain chemical substances and tensile stresses. The causes are usually electrochemical in nature.

	Acids			Bases		Hydrocarbons			Fats, oil
	Weak	Strong	Oxydizing	Weak	Strong	Aliphatic	Chlorinated	Aeromatic	
High-density polyethylene	+	+	−	+	+	+	−	○	+
Polypropylene	+	−	−	+	+	+	−	○	+
PVC-U (rigid)	+	○	−	+	+	+	−	−	+
PVC-P (flexible)	+	+	−	+	○	−	−	−	○
Polymethylmethacrylate	+	○	○	+	○	+	−	−	+
Polystyrene	+	○	−	+	+	○	−	−	+
Polytetrafluoroethylene	+	+	+	+	+	+	+	+	+
Polyamide (nylon)	−	−	−	+	○	+	○	+	+
Polycarbonate	+	+	○	−	−	+	−	−	+

+ = resistant ○ = conditionally resistant − = not resistant

(*a*)

	Acids			Bases		Hydrocarbons			Fats, oil
	Weak	Strong	Oxydizing	Weak	Strong	Aliphatic	Chlorinated	Aeromatic	
UP Resins	○	−	−	+	+	+	○	−	+
EP Resins	+	−	−	+	○	+	○	+	+
Crosslinked polyurethanes	○	−	○	+	−	+	○	+	○
Flexible polyurethanes	+	−	−	○	○	+	−	−	+
Silicone resins	+	−	−	+	+	○	−	○	+
Urea molding compounds	○	−	−	+	○	+	+	+	+
Melamine molding compounds	○	−	−	+	−	+	+	+	+
Phenolic molding compounds	+	−	−	+	−	+	+	+	+

+ = resistant ○ = conditionally resistant − = not resistant

(*b*)

Figure 4.7.1 (a) Resistance of some thermoplastics at room temperature. (b) Resistance of some thermosets at room temperature.

This fracture phenomenon also occurs in plastics in the simultaneous presence of internal *orientation-induced* or cooling-induced stresses (process related) and/or external tensile stresses and certain liquids or vapors. Since in this case, however, purely physical processes (wetting, diffusion, swelling) are usually involved, this phenomenon is called *stress cracking* (there is no real corrosion process). Almost all thermoplastics exhibit stress cracking when exposed to certain substances, with the following factors increasing the effect further:

- higher tensile stress,
- longer exposure,
- higher temperature,
- higher crystallinity,
- lower molecular weight.

Stress cracks often occur in use if molded parts are subsequently bonded with unsuitable adhesives, printed with solvent-containing inks, or come into contact with stress crack-initiating substances. Stress cracks are especially visible in unfilled, crystal clear plastics such as PS, SAN, PMMA, and PC.

4.7.3 Diffusion and Permeation

Because of their low density, plastics are relatively permeable with respect to gases and liquids. Mass transport—also call *permeation*—consists of the following process steps:

- *adsorption* of the diffusing substance at the interface,
- *diffusion* of the attacking substance through the material,
- *desorption* of the diffusing substance at the opposite surface, and removal of the diffused substance.

The permeation behavior with respect to low-molecular-weight substances is of particular interest when plastics are used as packaging materials, fuel containers, pipes for in-floor heating, and the like.

In food packaging applications, the primary requirement is to have as little oxygen and/or water vapor as possible penetrate the packaging in order to achieve the longest possible shelf life. Permeability values compiled from the literature (Fig. 4.7.2) permit a rough estimate to be made of the amount of gas or liquid passing through a film in a given period of time. This *permeation rate* is generally inversely proportional to the film thickness, an indication that diffusion processes are the determining factor. The permeation behavior can be decisively improved through the use of custom-engineered materials (packaging films with up to seven layers) or through chemical surface modification (e.g., *sulfonation* or *fluorination* of HDPE fuel tanks).

4.7.4 Weathering

The following are important factors that act on a product during weathering and result in changes to various properties:

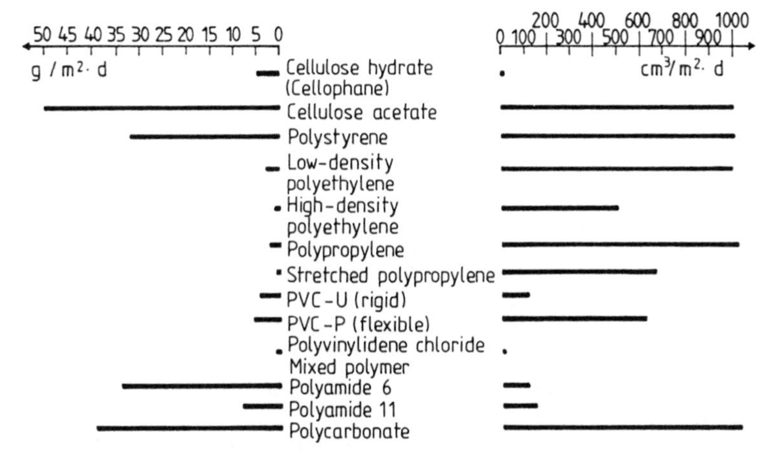

Figure 4.7.2 Water vapor permeability as per DIN 53122 and oxygen permeability as per DIN 53380 (temperature: 20°C, film thickness: 40 μm).

- *Radiation*: The entire radiation impinging on the surface of a body consists of the incident radiation coming directly from the sun, for instance, and the diffuse radiation reflected from the atmosphere or the environment, and is called the *global radiation*. Only a few specific molecules can be split directly by the absorbed light. In the presence of oxygen in the atmosphere, *oxidative processes*, which are initiated by light, assume the primary role in degradation.
- *Temperature*: Black plastic parts reach temperatures of more than 78°C under natural weathering conditions in the open; in the interior of an automobile, these temperatures can be exceeded considerably. In the course of some chemical (degradation) reactions, a temperature increase of 10 K leads to a doubling of the *reaction rate*. Accordingly, the temperature is of great significance.
- *Moisture*: Moisture plays a role above all with hydrolysis-sensitive plastics or additives.

In nature, the situation is aggravated in that the above-mentioned stresses do not occur individually, but rather in complex combinations, and can even be amplified through synergisms or an additional mechanical stress.

As soon as plastics were developed, it was necessary to prevent or at least delay the damage caused by climatic conditions through the addition of suitable substances (*UV absorbers, stabilizers*). The action of these additives is based either on a reduction of the light radiation that is damaging to a specific plastic (filter effect) or on a specific interruption of the chemical degradation reaction.

To evaluate the weathering resistance, weathering tests are conducted in the open atmosphere or accelerated tests are run in instruments with simulated global radiation at a higher intensity. During the weathering, the properties required for a specific application, such as color, gloss, mechanical behavior, and the like, are continuously monitored in order to obtain information on changes to the material and, finally, on the life expectancy of the product. Direct correlation of artificial weathering with natural weathering is only conditionally possible at present.

4.7.5 Biological Behavior

Natural polymers such as cellulose, starches, and proteins are subject to severe attack by *microorganisms* and can be degraded by them. In contrast, synthetic polymers as such can serve only with difficulty or not at all as a source of nutrition for microorganisms. The majority of low-molecular-weight organic products employed as processing aids, however, and especially the *plasticizers*, can be attacked.

Accordingly, the biologically based deterioration of a material, above all plasticized PVC (e.g., roofing and ceiling strips), can be observed when the plasticizer can be consumed as a source of nutrition by microorganisms. This leads to embrittlement of the material, with a resultant loss in functionality. Through the use of so-called *biostabilizers* (biocides), the start of biological degradation of plasticizers can be delayed considerably. Self-dissolving threads and temporary prostheses represent an important field of application for plastics in surgery, where intentional biological degradation is desired. Bone pins and plates, for instance, of polymers that gradually dissolve under the action of the body's enzymes, help to eliminate the need for a second operation to remove these aids and are at present undergoing clinical tests. Starch-filled polyethylene films are being tested as ripening accelerators and packaging material. Under composting conditions, the starch is eaten by bacteria, thereby increasing the specific surface area of the polymer considerably, which can lead to accelerated oxidative degradation of the polyethylene itself.

In addition, polymers are being developed that can be degraded directly by bacteria under composting conditions. Because of their very high price, these polymers are not suitable at present for large-scale commercial use, such as, for instance, in agriculture or packaging.

4.7.6 Flammability

All organic substances are based in carbon and thus combustible. Even the high-molecular-weight substances—regardless of whether they occur in nature in wood or as synthetically prepared plastics—are not an exception. With the introduction of a sufficient amount of energy, they will thus burn whether *flame-retardants* are present or not. Experience has shown that, in comparison with other combustible building materials such as wood, for instance, plastics do not present an increased risk of fire. For this reason, only suitable construction materials are considered in technical specifications; plastics are given neither positive nor negative emphasis.

In order for a fire to occur at all, it is necessary for three components to be present: combustible material, oxygen, and energy. The interaction of these three components results in the exceptional complexity of every fire.

As a fundamental standard for testing the flammability of building materials and components, DIN 41 02 has been introduced throughout Germany for the purpose of building inspections. According to this standard, materials are classified as "noncombustible" in Class A and "combustible" in Class B. As a rule, plastics are assigned to Class B. Within Class B, the classification ranges from B1—poorly flammable—through B2—normally flammable—to B3—readily flammable. By incorporating flame-retardants, the *flammability* can be influenced. The electrical (cable insulation, TV and audio equipment, telephone systems), the automotive (interior components), construction (floor coverings, plumbing), and furniture industries provide the primary fields of application for flame-retardant plastics.

4.8 Serviceability and Quality Assurance

The *practical behavior* of plastics parts is a function of the starting materials, the production conditions, the engineering design, and, last but not least, the combined stresses to be withstood.

The *serviceability* of a molded part, i.e., the retention of *functional integrity* during the intended service life, can often be specified to only a limited degree. Reasons for this may include inadequate description of the functional requirements, inexact knowledge of the combined stresses, as well as no possibility for design calculations.

For a multitude of applications, evaluation according to *quality criteria*—material, product inspection, production conditions—is sufficient. To assure quality, *quality control* (Fig. 4.8.1) has been introduced. Here, the requirements are established on the basis of *molded part tests* under near-practical conditions.

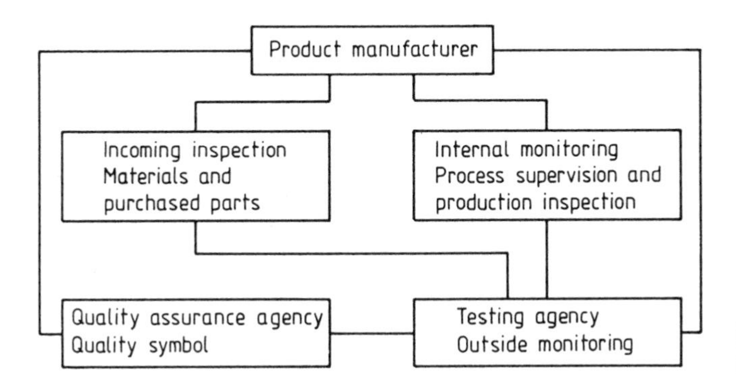

Figure 4.8.1 Schematic for quality assurance.

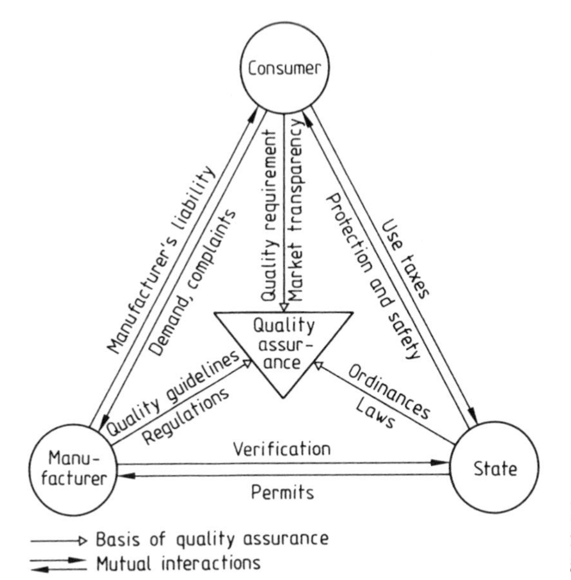

Figure 4.8.2 Quality assurance in relation to the consumer, manufacturer, and state [according to Krüger (1985)].

Effective *quality assurance* requires the cooperation of all participants (Fig. 4.8.2), just as the advantages of product quality control are recognized by all participants—state and business, manufacturers, retailers and consumers. Through quality control, the manufacturer can optimize production while reducing the reject rate and warranty claims. It represents an important aspect of *product liability*. For the retail trade, these measures promote activity and provide order in the market. Lastly, they serve to protect and reassure the consumer. Overall, the economic benefit of quality control is undisputed.

Bibliography for Chapter 4

Carlowitz, B.: Tabellarische Übersicht über die Prüfung von Kunststoffen, Kunststoff-Verlag, Isernhagen, 1986

Bargel, H.J., Schulze, G.: Werkstoffkunde, VDI-Verlag GmbH, Düsseldorf, 1980

Brown, R. P.: Taschenbuch Kunststoff-Prüftechnik, Carl Hanser Verlag, München, Wien, 1984

Domininghaus, H.: Die Kunststoffe und ihre Eigenschaften, VDI-Verlag GmbH, Düsseldorf, 1986

Ehrenstein, G.W.: Polymer-Werkstoffe, Struktur und mechanisches Verhalten, Carl Hanser Verlag, München, Wien, 1978

Hellerich, W., Harsch, G.C., Haenle, S.: Werkstoff-Führer Kunststoffe, 6th ed., Carl Hanser Verlag, München, Wien, 1992

Laeis, W.: Einführung in die Werkstoffkunde der Kunststoffe, Carl Hanser Verlag, München, Wien, 1972

Menges, G.: Werkstoffkunde Kunststoffe, 3rd ed., Carl Hanser Verlag, München, Wien, 1990

Oberbach, K.: Kunststoff-Kennwerte für Konstrukteure, Carl Hanser Verlag, München, Wien, 1975

Saechtling, H.: Baustofflehre Kunststoffe, Carl Hanser Verlag, München, Wien, 1975

Saechtling, H.: Baustofflehre Kunststoffe, Carl Hanser Verlag, München, Wien, 1975

Schreyer, G.: Konstruieren mit Kunststoffen, Carl Hanser Verlag, München, Wien, 1972

Bibliography for Chapter 4.1

Schmachtenberg, E.: Die mechanischen Eigenschaften nichtlinear viskoelastischer Werkstoffe, Doctor theses at RWTH Aachen, 1985

DIN 7724: Gruppierung hochpolymerer Werkstoffe aufgrund der Temperaturabhängigkeit ihres mechanischen Verhaltens; Grundlagen, Gruppierung, Begriffe

DIN Taschenbuch Nr. 18: Kunststoffe 1; Mechanisch, thermische und elektrische Eigenschaften, Prüfnormen, Beuth Verlag GmbH, Berlin, Köln

DIN Taschenbuch Nr. 48: Kunststoffe 2; Chemische und optische Gebrauchseigenschaften, Verarbeitungseigenschaften, Prüfnormen, Beuth Verlag GmbH, Berlin, Köln

Bibliography for Chapter 4.2

Thebing, U.: Beitrag zur Dimensionierung von GF-UP unter wechselnden Beanspruchungen, Doctor theses at RWTH Aachen, 1979

DIN 53373: Durchstoßversuch mit elektronischer Meßwerterfassung

DIN 53398: Biegeschwellversuch

DIN 53421: Druckversuch an harten Schaumstoffen

DIN 53423: Biegeversuch an harten Schaumstoffen

DIN 53430: Zugversuch an harten Schaumstoffen

DIN 53435: Biegeversuch und Schlagbiegeversuch an Dynstat-Probekörpern

DIN 53441: Spannungsrelaxationsversuch
DIN 53442: Dauerschwingversuch im Biegebereich an flachen Probekörpern
DIN 53443: Stoßversuch
DIN 53444: Zeitstand-Zugversuch
DIN 53448: Schlagzugversuch
DIN 53452: Biegeversuch
DIN 53453: Schlagbiegeversuch
DIN 53454: Druckversuch
DIN 53455: Zugversuch
DIN 53457: Bestimmung des Elastizitätsmoduls im Zug-, Druck- und Biegeversuch

Bibliography for Chapter 4.3

Racké, H.: Verhalten von Kunststoff-Oberflächen bei berührender und mechanischer Beanspruchung, part 2, Schreyer, G., see Bibliography for Chapter 4
Müller, Kl.: Härteprüfungen von Kunststoffen, Beschichtungen und Gummi, in Weiler, W.: Härteprüfung an Metallen und Kunststoffen, Expert Verlag, Sindelfingen, 1985
DIN 50133: Härteprüfungen nach Vickers (Metalle)
DIN 50320: Verschleiß, Begriffe
DIN 50321: Verschleiß, Meßgrößen
DIN 53375: Bestimmung des Reibungsverhaltens
DIN 53754: Bestimmung des Abriebs nach dem Reibradverfahren
DIN EN 59: Bestimmung der Härte mit dem Barcol-Härteprüfgerät
DIN 53456: Härteprüfungen durch Eindruckversuch
DIN 53505: Härteprüfung nach Shore A und D
DIN 53512: Bestimmung der Rückprall-Elastizität (Kautschuk)
DIN 53573: Bestimmung der Rückprall-Elastizität (Schaumstoffe)
DIN 53519: Bestimmung der Kugeldruckhärte Internationaler Gummihärtegrad (IRHD)

Bibliography for Chapter 4.4

Bednarz, J.: Kunststoffe in der Elektrotechnik und Elektronik, Verlag W. Kohlhammer, Stuttgart, 1988
DIN VDE 0303: Verfahren zur Bestimmung der Vergleichszahl und Prüfzahl der Kriechwegbildung auf festen isolierenden Werkstoffen unter feuchten Bedingungen
DIN VDE 0303: Durchschlagspannung, Durchschlagfestigkeit, part 2
DIN VDE 0303: Bestimmung der dielektrischen Eigenschaften, part 4
DIN VDE 0303: Bestimmung der Lichtbogenfestigkeit, part 5
DIN VDE 0303: Beurteilung des elektrostatischen Verhaltens, part 8
DIN VDE 0303: Hochspannungs-Kriechstromfestigkeit, part 10
DIN VDE 0303: Dielektrische Eigenschaften fester Isolierstoffe im Frequenzbereich von 8, 2 bis 12,5 GHz, part 13
DIN VDE 0303: Elektrostatisches Verhalten von dünnen Folien und Bändern; Prüfung der Aufladbarkeit, part 14
DIN VDE 53483: Bestimmung der dielektrischen Eigenschaften, Begriffe, Allgemeine Angaben
DIN VDE 53484: Bestimmung der Lichtbogenfestigkeit

Bibliography for Chapter 4.5

DIN 53491: Bestimmung der Brechungszahl und Dispersion
DIN 67530: Reflektometer als Hilfsmittel zur Glanzbeurteilung

DIN 5033: Begriffe der Farbmetrik, Meßverfahren, Meßbedingungen
DIN 5036: Strahlungsphysikalische und lichttechnische Eigenschaften von Materialien
N.N.: Farbmetrik für Prüfung und Farbrezepturberechnung beim Einfärben von Kunststoffen, VDI-Verlag GmbH, Düsseldorf, 1981

Bibliography for Chapter 4.6

Gösele, K.: Schall, Wärme, Feuchte, Schüle, W. Bauverlag GmbH, Wiesbaden, Berlin, 1983
N.N.: Ultraschall in der Kunststoff-Fügetechnik, Herfurth GmbH, Hamburg, 1986
DIN 1320: Akustik, Grundbegriffe
DIN 1332: Akustik, Formelzeichen
DIN 45630: Grundlagen der Schallmessung, Physikalische und subjektive Größen von Schall
DIN 53445: Prüfung von Kunststoffen, Torsionsschwingversuch

Bibliography for Chapter 4.7

Doležel, B.: Die Beständigkeit von Kunststoffen und Gummi, Carl Hanser Verlag, München, Wien, 1978
Troitzsch, J.: Brandverhalten von Kunststoffen, Carl Hanser Verlag, München, Wien, 1982
DIN 4102: Brandverhalten von Baustoffen und Bauteilen
DIN 53380: Bestimmung der Gasdurchlässigkeit
DIN 53381: Bestimmung der Thermostabilität
DIN 53383: Prüfung der Oxidationsstabilität
DIN 53386: Bewitterung im Freien
DIN 53387: Bewitterung in Geräten
DIN 53388: Belichtung im Naturversuch unter Fensterglas
DIN 53393: Prüfung von glasfaserverstärkten Kunststoffen; Verhalten bei Einwirkung von Chemikalien
DIN 53449: Beurteilung der Spannungsrißbildung
DIN 53476: Bestimmung des Verhaltens gegen Flüssigkeiten
DIN 53739: Einfluß von Pilzen und Bakterien
DIN 53755: Lagerungsversuch bei thermischer und äußerer mechanischer Beanspruchung
DIN 53756: Lagerungsversuch bei chemischer Beanspruchung
DIN 16888: Bewertung der chemischen Widerstandsfähigkeit von Rohren aus Thermoplasten
DIN 16889: Bestimmung der chemischen Resistenzfaktoren an Rohren aus Thermoplasten

Bibliography for Chapter 4.8

Krüger, E. J: Aufbereitung von PVC. Doctor theses at RWTH Aachen, 1985
Masing, W.: Handbuch der Qualitätssicherung, Carl Hanser Verlag, München, Wien, 1981
Oberbach, K., Müller, W.: Prüfung von Kunststoff-Formteilen, 2nd ed., Carl Hanser Verlag, München, Wien, 1988
DIN 18200: Überwachung (Güteüberwachung) von Baustoffen, Bauteilen und Bauarten; Allgemeine Grundsätze
DIN 53760; Prüfung von Kunststoff-Fertigteilen; Prüfmöglichkeiten, Prüfkriterien
DIN 53350: Begriffe der Qualitätssicherung und Statistik; Begriffe der Qualitätssicherung, Grundbegriffe
DIN 66050: Gebrauchstauglichkeit

5 Compounding of Plastics

5.1 Introduction

The term *compounding* describes all process steps that occur between the synthesis of plastics resin and its forming in the processing machine. The transition from compounding to processing can be continuous. Compounding thus includes the steps of conveying, metering, mixing, pelletizing, and sometimes even storage.

The necessity for compounding results to a large degree from the fact that plastics cannot be processed directly after polymerization, i.e., without intermediate steps, or exhibit an inadequate property profile. Consequently, additives must be mixed with the base resin to permit the reliable production of parts on an industrial scale, keeping in mind that the product's properties must be matched exactly to the requirements associated with the future field of application.

Thus, a thermal stabilizer permits polymers to withstand without degradation the temperature level necessary for processing. For this reason as well, additional processing aids are added that, for instance, accelerate melting, influence the melt viscosity, or reduce the residence time of the material in the processing machine by preventing the resin from adhering to metallic parts. Other additives improve mechanical properties to such an extent that plastics can compete with traditional metals. Incorporation of fibers into the polymer matrix is one example that can be mentioned here. Through the use of plasticizers, polymeric materials that are inherently hard and brittle can be made flexible and expandable, as a result of which they can be employed for entirely new applications.

Even optical properties can be varied over a wide range through the use of additives. Pigments can be used to achieve specific gloss effects or produce plastic products in any desired color. Fillers are gaining in importance not only for their volume-increasing aspect but also for their effects on processibility and material properties. Another important aspect of compounding is that many plastics are initially produced in a form that is unsuitable for processing, e.g., a fine powder. In one of the compounding steps, *pelletizing*, the material is plasticated, shaped, and cut, yielding a plastic pellet with defined dimensions that exhibits better feed properties in the processing machine.

As mentioned at the beginning, every plastic must in principle be compounded. As a rule, however, this normally takes place at the material supplier so that the processor receives a ready-to-process compound. Of the thermoplastics, only polyvinyl chloride (PVC) is compounded on a large scale at the processor. (Many rubber processors also compound their own formulations.)

There are two main reasons why PVC is compounded at processors. On the one hand, no other polymer is as compatible with such a large number of additives. As a result, PVC can be formulated to range from flexible and elastic to highly impact-resistant. Its field of application extends from phonograph records to water and gas pipes, from packaging films and floor coverings to window profiles or wire coatings, to give only a few examples of applications. With PVC, the processor is able to provide a formulation tailored specifically to the requirements of his products. The material supplier cannot be expected to supply such a large number of formulations. The other reason is that PVC is a commodity resin. Its low price and the universal applicability discussed above have lead to a continuous increase in world production of PVC (14.2 million t in 1985, 17.7 million t in 1989, and an estimated 18.1 million t in 1990). This has resulted in large production facilities with correspondingly large outputs of PVC per year. The cost savings associated with in-house compounding then justify for these companies the high capital investments for compounding machinery (e.g., mixers).

5.2 Compounding Machines

The most important compounding machines will be presented in this section. For the above-mentioned reasons, this discussion will be limited to PVC compounding, even though many of the units can be employed for other plastics.

Storage and Conveying

Powdered PVC is generally kept in large storage silos that are usually filled from rail cars or trucks. Sacks of material are employed only for very small amounts or in an emergency when the silos cannot be used (Fig. 5.2.1).

As a rule, the silos are connected to the mixing and metering devices via pneumatic conveyors. All conveying equipment must be designed to high engineering standards, since the development of dust, for instance, can lead to dangerous conditions, and reliable removal of the often fine powder from the silos must be assured. The *conveying systems* are designed as either *pressure conveyors* or *vacuum conveyors*. In occasional instances, even pressure/vacuum combination systems are employed. Mechanical conveyors such as feed screws, conveyor belts, or vibrating chutes are utilized only for special applications.

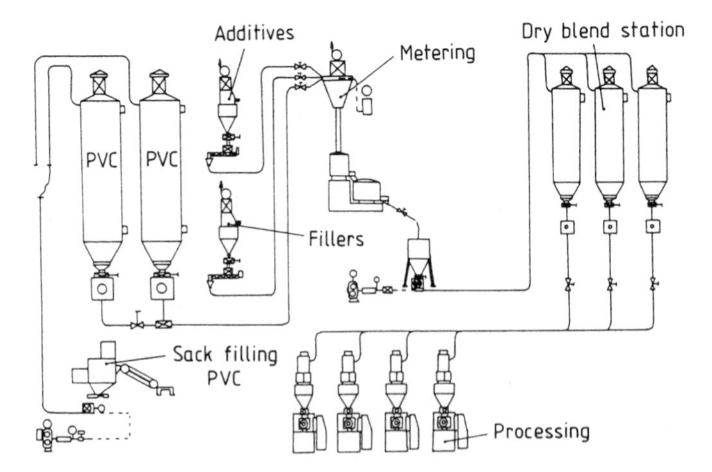

Figure 5.2.1 Schematic illustration of a discontinuous PVC compounding system.

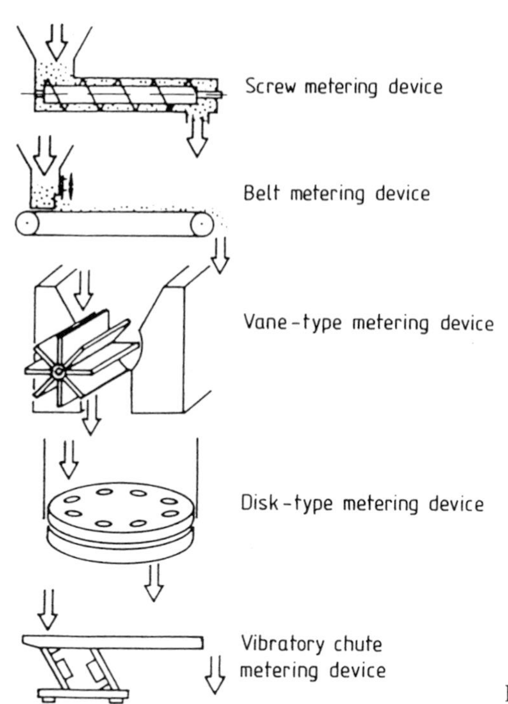

Figure 5.2.2 Volumetric metering systems.

Metering

Metering can be classified as gravimetric or volumetric. With regard to accuracy and the possibility of using automation, gravimetric metering, i.e., weighing of the individual components, is to be preferred. Such systems, however, have a relatively high price.

While *volumetric metering equipment* is less expensive to purchase, the metering accuracy is affected by a multitude of factors, e.g., the bulk density or particle size, which can vary from batch to batch. A number of volumetric metering systems are shown in Fig. 5.2.2.

With the screw-type metering device, the bulk material is conveyed by a screwlike auger mounted in a barrel. This system can be employed only when the material to be conveyed exhibits good flow properties.

The same applies to the belt-type metering device, where the material to be conveyed is discharged from a full hopper onto a conveyor belt. In this case, the volumetric flow is controlled by means of a slide gate.

In the wheel and plate metering devices, chambers are filled volumetrically and then emptied after a certain angle of rotation of the rotor or plate. The mass flow is varied via the speed of rotation.

Although vibrating chutes are not really volumetric metering devices, they are listed here for the sake of completeness. They operate on the principle of microdisplacement, which takes place when the chute is made to vibrate. This system is employed primarily with coarse and fine-grained materials.

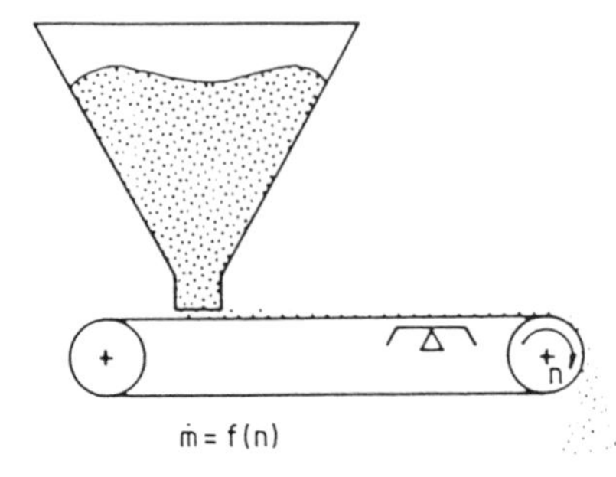

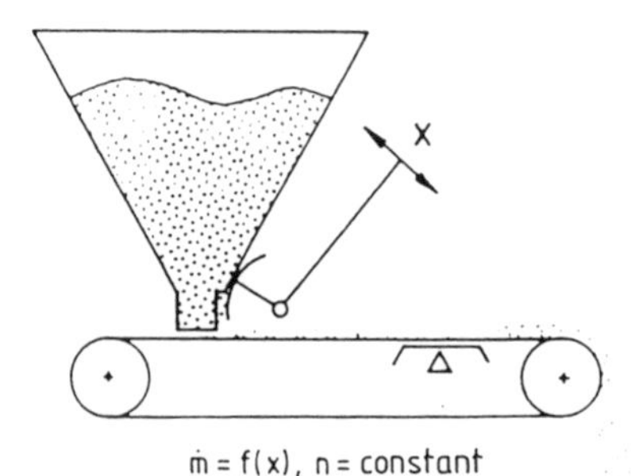

Figure 5.2.3 Gravimetric metering systems.

As mentioned above, limits are set to the metering accuracy for solids by batch-to-batch variations. With liquid additives, e.g., plasticizers, this disadvantage is not as pronounced, so that volumetric metering by means of pumps has become the accepted practice. With gravimetric metering, both a discontinuous and a continuous mode of operation are possible. In the discontinuous mode, the individual components are placed in a weighing container in succession and weighed additively. If for this mode of operation the necessary time is not available because of larger quantities of material, the additives and PVC must be weighed on separate scales at the same time and subsequently mixed, which naturally leads to a sudden increase in capital investment.

In contrast to this method, belt-type metering scales operate continuously. In this case, a weighing bridge with an exactly defined measurement is located beneath a conveyor belt (Fig. 5.2.3). The mass flow is measured continuously and serves as the control variable for the belt speed or degree of opening of the hopper.

Mixing

Preparation of a homogeneous mixture of PVC and additives sets demanding requirements on both the machinery and processing technology. The objective of mixing is to distribute the *additives* as uniformly as possible in the PVC base resin without degrading the PVC in the process. Two methods are distinguished: *cold mixing* at room temperature, where the individual components are only mixed together, and *hot mixing* at up to approximately 140°C, where absorption and diffusion processes also occur. A typical representative of the

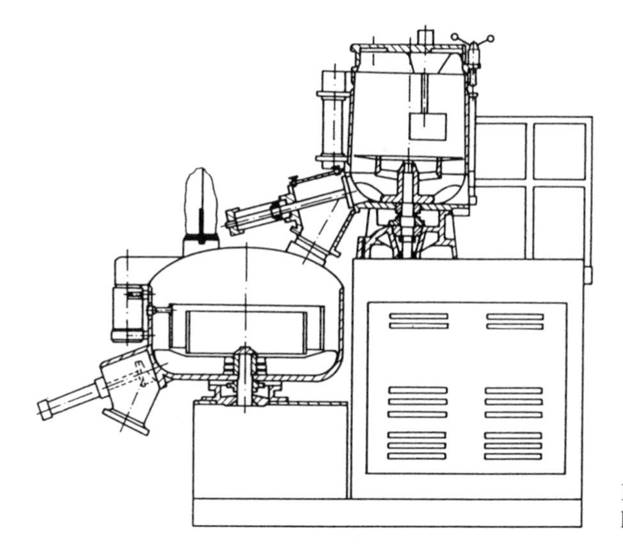

Figure 5.2.4 Combination heating–cooling mixer.

first group is the rotary tumbling mixer, in which the mixing process occurs solely under the influence of gravity, or the ribbon mixer, which mixes the material with the aid of a helical blade that is usually mounted in a horizontal cylinder. The second group is divided according to the manner in which heat is produced. On the one hand, the resin can be heated through contact with metal parts, irradiation or convection, e.g., in a plow-type mixer, or through dissipation of the mechanical drive energy introduced to the product in the form of frictional heat. Since high mixer speeds are required for this type of energy introduction, these mixers are called high-performance mixers or turbo mixers. This type of mixing permits heating of the material in a short period of time, and for this reason it has found wide acceptance in industry. Accordingly, it will be discussed in somewhat greater detail.

The mixing chambers, which are usually of double-wall construction for additional temperature control, contain ring- or propeller-shaped mixer blades, which generally rotate at a circumferential velocity of between 20 and 50 m/s. They are supposed to mix the individual components and simultaneously heat them by forcing the material against and up the chamber wall (Fig. 5.2.4). This results in a trombus-shaped motion through which the material once again reaches the bottom of the chamber after passing through the center. The powder is heated through intergranular friction and impingement of the particles against one another.

During the initial phase of the mixing process, the components are only intermingled homogeneously; the actual hot mixing begins only above 50°C. In the course of further heating, during which the plasticizing temperature of approximately 150°C should not be exceeded, solid additives are broken down in size and meltable additives are liquefied. The PVC pellet, which is not reduced in size during this phase, absorbs the liquid additives so that a dry powder, the so-called *dry blend*, results at the end of the hot mixing. This compound has a temperature of about 110–130°C and must be cooled in a cooling mixer for any necessary storage. This is achieved by letting the powder fall freely into the lower, water-cooled chamber of the cooling mixer, in which it is cooled to approximately 40°C as slowly rotating blades prevent the formation of lumps (Fig. 5.2.4). The dry blend produced in this manner can be processed without any further compounding steps, as usually happens with rigid PVC in counter-rotating twin-screw extruders; or it can be pelletized in a subsequent compounding process, as is usually the case with flexible PVC.

Plastication

Plastication of the resin during compounding is necessary if the material is to be pelletized, for instance, or processed into film or sheet on a calender. During this step, larger amounts of fillers can be added that would be uneconomical or impractical from a process standpoint in a hot mixer.

Roll Mills

The use of roll mills, where two heated rolls rotating in opposite directions draw the material out into thin sheets, is probably the oldest means for melting plastics. Heating results from contact with the hot rolls and through friction upon passing through the narrow gap between the rolls. The great advantage of this system is that it represents an open process to which additives or scrap material, e.g., edge trimmings from a subsequent processing step, can be easily reintroduced. In addition, the stock temperature can be held within narrow limits, which is an advantage that should not be underestimated with regard to temperature-sensitive PVC and when processing rubber formulations, since rubber formulations in particular must be homogenized with very high shear forces and at well-controlled temperatures in order to prevent premature cross-linking. Cleaning during batch changes also presents no problems, which is of particular importance with many PVC and rubber formulations.

Negative aspects include that the milling process is discontinuous and difficult to automate. Recently, however, a roll mill was developed that operates continuously with the aid of helical grooves machined into the roll surface.

Kneaders

Kneaders also compound plastics discontinuously. The material is kneaded between blades that rotate in a twin cylinder arrangement and is discharged in a plasticized condition after a certain period of time (Fig. 5.2.5).

A modified version of this equipment operates continuously. The kneading blades are designed as rotors on which several screw flights at the feed end convey material to the actual kneading elements. Connected to this kneading zone is the so-called pumping zone, which discharges the plasticized material (Fig. 5.2.6).

Planetary Roller Extruders

The planetary roller extruder is based on the principle of the single-screw extruder; it differs, however, in that the screw is designed as a spindle about which several planetary spindles rotate (Fig. 5.2.6a). With this arrangement, it is possible to shear and mix the material

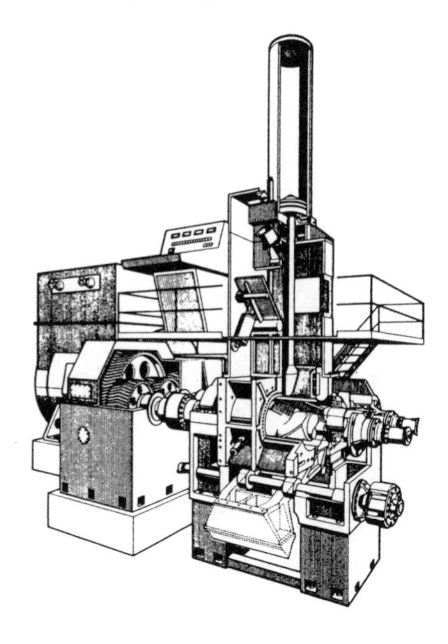

Figure 5.2.5 Discontinuous internal mixer.

intensely, since new surfaces are constantly being formed. As a result of the thin material films and the very large surface areas of the central and planetary spindles, it is possible to control the melt temperature exactly.

Plasticators

With the aid of a feed screw, the raw material is fed to a conical rotor inside a conical housing (Fig. 5.6b). The gap between the rotor and the housing can be adjusted by shifting the rotor with respect to the housing. The good homogeneity of the plasticated material results from the powdered material first being sintered together in the gap and subsequently being plasticated. The layer of material is expanded by the cone and continuously divided and layered by helically arranged grooves or flights. Good temperature control is also

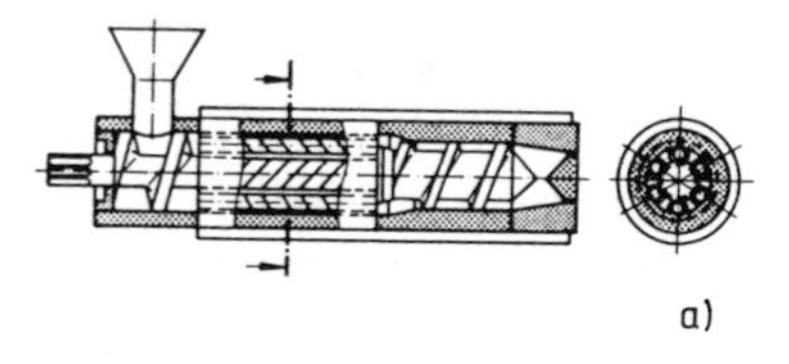

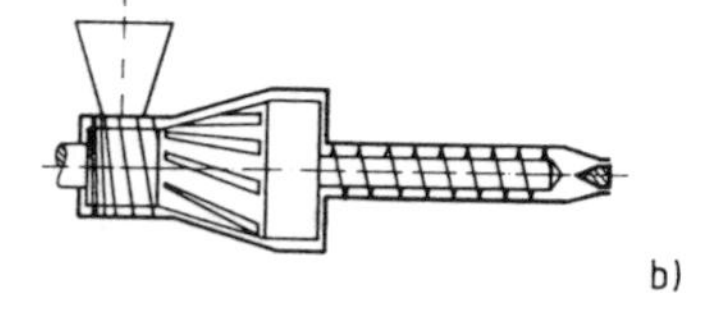

Figure 5.2.6 Sketches illustrating the principle of a planetary roller extruder (a) and a plastificator (b).

achieved here through the large surface area. The plasticized PVC is then fed to the downstream equipment by a discharge screw.

Twin-Screw Extruders

Both co-rotating and counter-rotating twin-screw extruders are employed for compounding. For this application, they have a modified screw geometry and are equipped in part also with kneading and mixing elements (see also Chapter 6).

Pelletizing

Pelletizing is accomplished in one of two manners: *hot pelletizing* or *cold pelletizing*. Depending on the method, the resulting pellets are lens-shaped, spherical, cylindrical, or diced.

In the cold pelletizing process, strands, strips, or sheets are formed by a tool and then cut into the desired shape by a rotating knife once they have solidified after cooling. In contrast, with hot pelletizing the plasticated material is extruded through a die plate, and the exiting thermoplastic strands are cut by a knife rotating across the face of the die plate. Cooling takes place only after the pelletizing step, e.g., with the aid of water in underwater pelletizing.

5.3 Additives

In this section, the discussion will focus on additives, with those used for PVC by way of example.

Stabilizers

PVC releases hydrogen chloride at temperatures above 100°C. The result is a darkening or a blackening and simultaneous embrittlement of the material. Since, however, the processing temperatures lie between 160 and 210°C, a thermal stabilizer is absolutely necessary. Such a stabilizer must satisfy several requirements.

It must absorb and neutralize the HCl released so that machinery and equipment are not attacked, and it must prevent oxidation and other radical-induced processes. It must further prevent the formation of double bonds, because these result in the above-mentioned discoloration of the material.

In addition, it should be mentioned here that PVC can also decompose upon exposure to light (UV radiation) and air (oxygen). The material thus must be stabilized against these.

The most important examples of thermal stabilizers are the barium-, calcium-, zinc-, lead-, and metal-free stabilizers as well as organotin compounds. They differ in the way they function and their effectiveness. Mixtures of the mentioned stabilizers are also employed very often, since the effect of a mixture is greater than the sum of the effects of the individual components ("synergy effect").

The selection depends on the required properties of the final product, economic aspects, or, as in products for food contact applications, is specified by law.

Plasticizers

PVC at room temperature is a hard resin that does not become flexible until approximately 70°C. By incorporating plasticizers, this limit can be brought down to lower temperature values. Thus, depending on the amount and type of plasticizer added, it is possible to produce a soft, flexible, and expandable material down to −20°C. This can be achieved on the one hand through so-called *internal* plasticizing, whereby comonomers containing no chlorine and having a different polarity than PVC are incorporated into the PVC molecule during polymerization. The advantage of this plasticizing is that the resin remains very stable over a long period of time and the plasticizer cannot diffuse away. It does present disadvantages, however, with regard to flexibility and strength at low temperatures.

For low-temperature applications, *external* plasticizers that physically bond via dipoles have proven successful. External plasticizing predominates in actual practice, since the processor can achieve the desired product properties via the type and amount of plasticizer. It is for this reason that an almost unlimited variety of plasticizers is offered for the various applications.

With a market share of over 60% of the overall plasticizer market, the phthalates occupy first place among all plasticizers employed because they combine a low price with good properties at the same time. The two most important representatives of this group are DOP (dioctyl phthalate) and DIOP (di-iso-octyl phthalate).

In addition to the phthalates, epoxies, aliphatic esters of dicarboxylic acid, phosphates, and polyesters are employed, to name only a few.

Impact Modifiers

The impact toughness, which is quite low in comparison to other polymers, can be increased by incorporating elastomeric components into the PVC matrix. The action of these high-impact modifiers is based on the fact that these polymeric materials cannot form a homogeneous mixture with the PVC, so that a heterogeneous mixture with separate phases results. It is for this reason that not only the chemical composition is important for the achievable results, but also the morphology, i.e., in which form the additives are present.

Impact modification can be accomplished in two manners, namely through graft polymerization of the elastomer component with the vinyl chloride monomer or through admixture of the modifier during compounding. Common modifiers include EVA (ethylene vinyl acetate), PAE (polyacrylic ester), PEC (chlorinated polyethylene), or ABS (acrylonitryl-butadiene- styrene).

Fillers

Organic or inorganic additives that are usually in a solid form and which differ significantly in their structure and composition from the resin are called *fillers*. As the name already indicates, these materials were originally used only to increase volume, i.e., reduce the costs of the finished parts. Now this has changed. Today fillers are also employed specifically to improve properties and processibility.

In general, fillers increase the stiffness and reduce the modulus of elasticity and/or tensile strength. Commonly employed fillers include stone flour (chalk), paper fibers, wood flour, carbon black, graphite, talc, or metal oxides. Glass fibers can also be counted among the fillers. They offer more than any other filler the opportunity to influence the end-use properties of a product. They are used to only a very limited degree, however, with PVC.

Lubricants

It is only the use of lubricants that has made processing of rigid PVC possible today. In order to avoid unnecessarily high stresses on the thermally sensitive PVC, *internal lubricants* are employed to improve the flow characteristics of the melt so that excessive frictional heat is not generated. *External lubricants* assure that the PVC melt does not adhere to hot components such as barrel, screw, or rolls, and that cold assemblies such as calibrating devices or chill rolls do not lead to uncontrolled residence times. Examples of lubricants include paraffin oils, calcium stearate, and stearic acids. It should be mentioned in conclusion that it is possible to discuss only the most important additives here. As a rule, a PVC formulation involves a multitude of individual components, some of which are present in very low concentrations but which nevertheless have a decisive influence on the material behavior as a whole. These substances include, among others, colorants, pigments, optical brighteners, flame retardants, UV absorbers, antistats, and release agents.

Bibliography for Chapter 5

Domininghaus, H.: Die Kunststoffe und ihre Eigenschaften, VDI-Verlag GmbH, Düsseldorf, 1986
Becker, W., Braun, D.: Kunststoff-Handbuch Polyvinylchlorid 2/1, edited by H.K. Felger, Carl Hanser Verlag, München, Wien, 1986
Becker, W., Braun, D.: Kunststoff-Handbuch Polyvinylchlorid 2/2, edited by H.K. Felger, Carl Hanser Verlag, München, Wien, 1986
N. N.: Granulieren von thermoplastischen Kunststoffen, VDI-Verlag GmbH, Düsseldorf, 1974
N. N.: Aufbereiten von PVC, VDI-Verlag GmbH, Düsseldorf, 1976
Kopsch, H.: Kalandertechnik, Carl Hanser Verlag, München, Wien, 1978

6 Plastics Processing Methods

6.1 Extrusion

Extrusion is the continuous production of a semi-finished article, pipe, or film/sheet from a plastics resin. Among other components, an extrusion systems consists of:

- extruder,
- die,
- calibration/cooling device,
- take-off,
- cutoff device,

as shown in Fig. 6.1.1 with a pipe extrusion line by way of example. The construction and operation of the individual system components are described in greater detail in the following.

6.1.1 The Extruder

The extruder is the heart of every extrusion system and must supply a homogeneous melt to the attached die in sufficient quantity at the necessary temperature and pressure. Different types of extruders are employed depending on the application. Fig. 6.1.2 presents an overview of the various types.

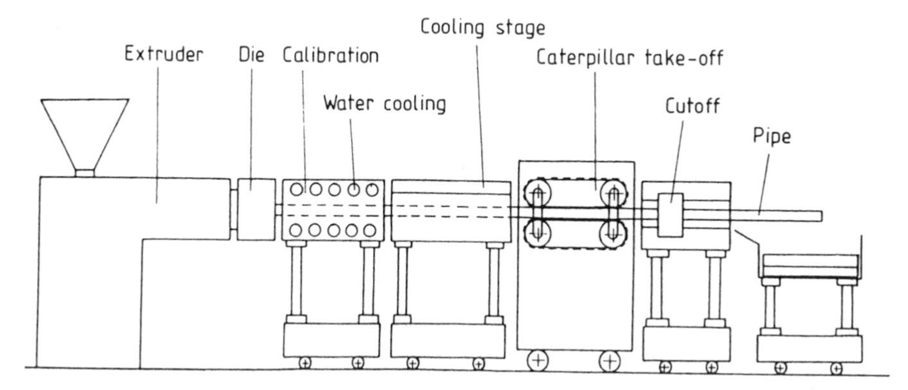

Figure 6.1.1 Pipe extrusion line (principle).

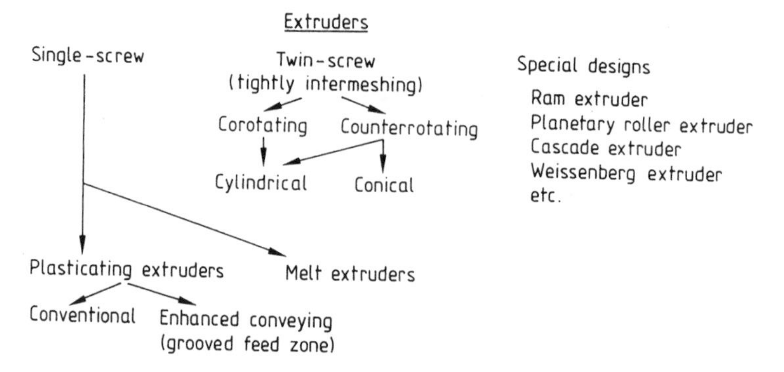

Figure 6.1.2 Classification of screw extruders.

An extruder normally consists of (Fig. 6.1.3):

- a machine base,
- a drive (motor, gearbox, thrust bearing),
- a plasticating unit [(screw(s), barrel, temperature control system],
- a control cabinet (control logic, control devices, power supply).

Today four different types of plasticating extruders are employed for the most part:

- single-screw extruders with a smooth feed zone and three-zone screw;
- single-screw extruders with a grooved feed zone and screw with mixing and, possibly, shearing sections;
- co-rotating twin-screw extruders;
- counter-rotating twin-screw extruders.

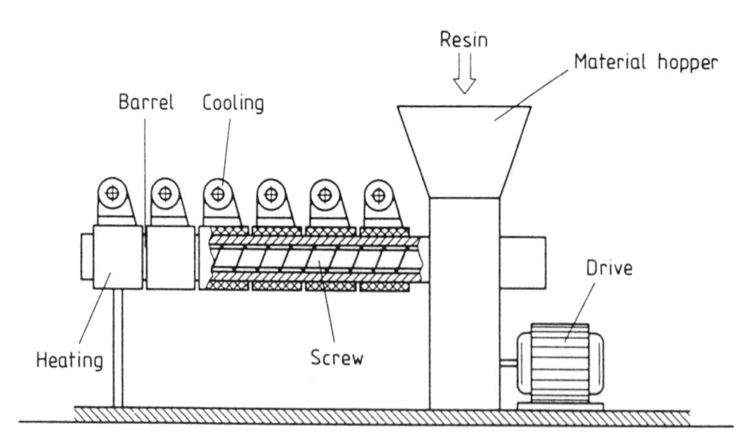

Figure 6.1.3 Extruder.

Function		Process characteristics	Engineering solution
Conveying	Solids	(Low) bulk density Low drag forces	Deeply cut flights Grooved feed zone
	Melt	Higher melt density Usually barrel wall adhesion	Shallower flights
Plasticating		Solids - melt mixture with poor heat conduction and increasing density	Decreasing flight depth Shear elements
Homogenizing		Inhomogeneous material and temperature distribution	Shallow-cut mixing zone Mixing section

Figure 6.1.4 Functions of a plasticating extruder.

6.1.1.1 The Single-Screw Extruder

A basic distinction is made between melt-conveying and plasticating extruders. The basic concept underlying both types of machines, i.e., controls, drive, gearbox, etc., is largely identical. Depending on the type of material (melt or pellets) to be conveyed, however, different specific screw geometries result for melt-conveying and plasticating extruders.

The function of a melt-conveying extruder is limited to conveying and homogenizing the melt. Since this type of extruder is encountered only seldom, it is not considered further in the following. All subsequent discussions deal with the plasticating extruder.

An extruder is usually characterized by the screw diameter and effective screw length, with the latter given as a multiple of the screw diameter. Example: 50 mm/25 D means a screw length of 50 mm × 25 = 1250 mm.

The functions of a plasticating extruder are summarized in Fig. 6.1.4. Conveying of the material in the extruder results from adhesion of the melt to the barrel wall and screw, which move relative to one another. This mechanism is called *drag flow*. The so-called drag flow G_s that results has superimposed on it a pressure flow G_p, which depends on the particular pressure gradient. This pressure gradient can assume positive or negative values—depending on the extruder concept.

The leakage flow G_l represents a third flow component. It describes the volumetric flow between the screw flight and barrel wall. Because of the narrow gap here, the material experiences a great deal of shear. Fig. 6.1.5 shows the three flow components.

The volumetric rate of discharge (output) G is thus given by:

$$G = G_s - G_p - G_L$$

or, if the leakage flow G_l can be neglected, then

$$G = G_s - G_p$$

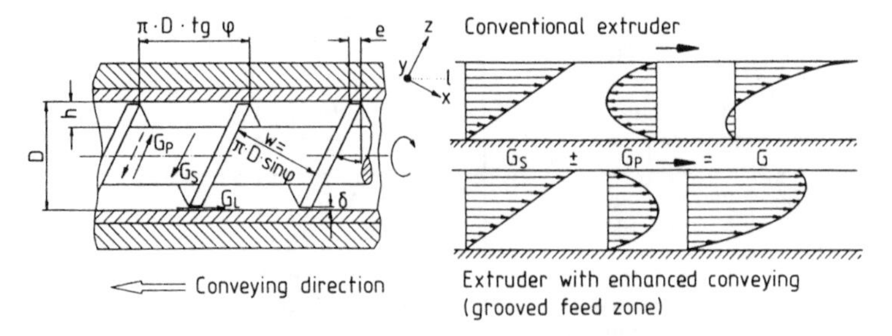

Figure 6.1.5 Velocity fields in a conventional extruder and an extruder with enhanced conveying (grooved feed zone); definition of screw geometry.

The dependence of these quantities on the geometry of the screw provides essential data for its design. Using the designations in Fig. 6.1.5 and the screw speed n, it follows that drag flow,

$$G_s = K_s \cdot h \cdot n$$

$$K_s = \frac{1}{2} D^2 \cdot \sin \varphi \cdot \cos \varphi$$

i.e.,

$$G_s \sim H \cdot n \cdot D^2$$

This relationship can be derived from the so-called two-plate model (see Fig. 3.2.1). Here the rotating surface of the screw corresponds to the moving plate in the model.

$$G_p = K_p \frac{p \cdot h^3}{\eta \cdot L}$$

$$K_p = \frac{1}{12} D \cdot \sin^2 \varphi$$

p	pressure
n	screw speed
η	viscosity
L	screw length

i.e.,

$$G_p \sim \frac{p \cdot h^3 \cdot D}{\eta \cdot L}$$

The graphic representation of the superposition of pressure flow and drag flow is also shown in Fig. 6.1.5 for the single-screw extruder with smooth feed zone and three-zone

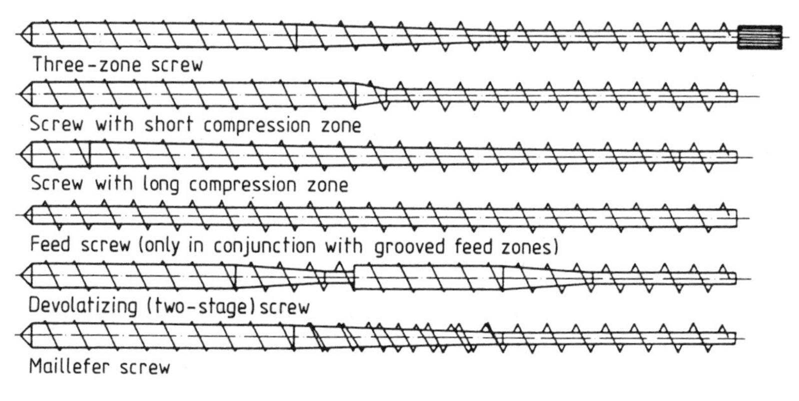

Figure 6.1.6 Common plasticating screws.

screw (conventional extruder) as well as for the single-screw extruder with grooved feed zone (enhanced conveying extruder). This is discussed in greater detail in the following.

6.1.1.2 The Screw

The screw is the heart of an extruder. Fig. 6.1.5 shows the characteristic dimensional designations.

The following designations are employed:

D	diameter	W	channel width	G_s	drag flow
φ	helix angle	e	land width	G_p	pressure flow
h	channel depth	δ	radial clearance	G_l	leakage flow

To satisfy the multiple requirements (feeding, conveying, plasticating, homogenizing), the geometry changes along the length. Fig. 6.1.6 shows a number of common screw geometries, with the so-called three-zone screw among the most common. For this reason, it will be discussed in greater detail.

As the name already indicates, it contains the following three zones:

- *Feed zone*: This pulls in the plastic pellets from the hopper and conveys them.
- *Compression zone*: This compresses the material conveyed from the feed zone and plasticates it.
- *Metering zone*: This homogenizes the melt and brings it to the desired temperature. In conventional extruders the pressure is also built up here at the same time. It thus also determines the output in such extruder systems.

With this screw geometry and any additional shearing and mixing sections that may be necessary for a given resin, almost all thermoplastics can be processed satisfactorily from a technical and economical standpoint.

Nevertheless, other designs are justified for specific applications. The short-compression screw, where the channel depth decreases from the deeply cut feed zone to the shallow metering zone in a very short compression zone, has proven useful for semi-crystalline thermoplastics.

For PVC there is an additional special design, the constant taper screw, where the root diameter increases uniformly over the full flighted length. Examples of these are also shown in Fig. 6.1.6. The specially designed extruders and screws can be considered a technically satisfactory solution if the following requirements are met:

- constant, pulsation-free conveying;
- thermally and compositionally homogeneous melt;
- operation within the limits for thermal, mechanical, and chemical material degradation.

For economic reasons, the above requirements, which must be satisfied in any case, must be combined with:

- high volumetric output at low specific operating costs.

Depending on the application and required output, the screw diameters common today vary from the 19-mm laboratory extruder to the special design with $D = 300$ mm. The length varies from 6 D for the melt-conveying extruder and rubber processing to approximately 25 D for plasticating extruders common in the market today. In addition to the extremely long vented extruders and those for production of plastics foams (approximately 32 to 35 D), conventional plasticating extruders with up to 30 D can also be found in industrial practice.

The classification of screws requires, besides diameter and length, information on the distribution of the functional zones and the compression ratio, which is defined as the ratio of the channel depth in the metering zone h_2 to that in the feed zone h_1. In order to guarantee filling of the screw, the compression ratio must correspond at least to the ratio of the bulk density of the solids in the feed zone to the density of the melt in the metering zone.

The two-stage (devolatilizing) screw also shown in Fig. 6.1.6 is employed in so-called *vented* extruders. These extruders have an opening to release gaseous components and entrapped air at the location where the channel depth makes an abrupt increase. The sudden enlargement in cross section results in a severe pressure drop, so that resin does not escape but volatile components can be removed by applying a vacuum.

The screws described here are used in two different extruder concepts, which differ in the design of the extruder barrel in the feed zone.

Conventional Single-Screw Plasticating Extruders

In what is called the conventional design for historical reasons, which is characterized by smooth barrel wall in the region of the feed zone, the metering zone is considered to determine the output. The resin that has been melted in the compression zone enters the metering zone, where it is homogenized and finally discharged through the attached extrusion die at the necessary pressure. In this concept, the metering zone serves to build up pressure. The variation in pressure over the screw length is directly dependent on, among other things, the back pressure at the die as well as the screw geometry (Fig. 6.1.7).

In the metering zone, the drag flow and pressure flow are superimposed as shown in Fig. 6.1.5. Because of the resistance at the die, this extruder concept exhibits a pressure flow

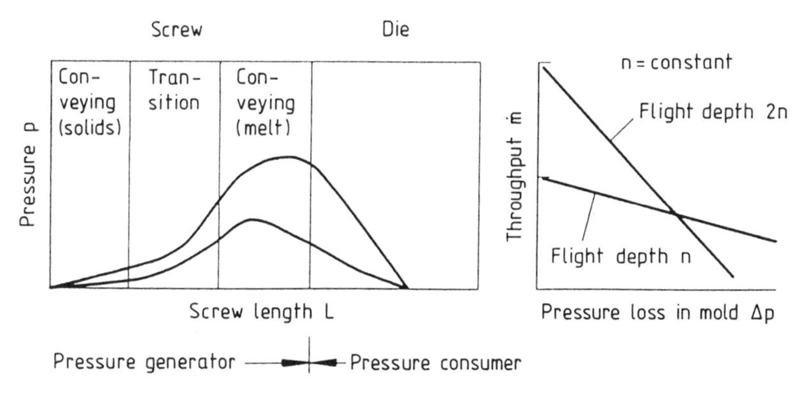

Figure 6.1.7 Pressure curve and throughput/die back-pressure curve for a conventional single-screw extruder.

opposite the drag flow and thus, with increasing channel depth, a greater influence of the pressure flow on the reduction in overall output—as can be seen from the descriptive equations—i.e., shallow-cut screws exhibit less of a back pressure dependence. This is also illustrated in Fig. 6.1.7.

The conveying mechanism in the solids region is determined by the Coulomb friction, which acts between the barrel wall and pellet as well as between the pellets themselves. Since the coefficient of friction decreases greatly as the pellets melt, leading to inadequate conveying, the feed zone is generally cooled to prevent melting of the pellets.

Nevertheless, the strong dependence of the output on the back pressure remains a serious deficiency of the conventional extruder. An improvement in this regard is provided by an extruder with enhanced conveying, the so-called *extruder with grooved feed zone*.

The Single-Screw Extruder with Grooved Feed Zone

To achieve high outputs independent of the back pressure, feed zones with axial grooves were developed. By interrupting the otherwise smooth barrel wall with regularly spaced axial grooves, movement of the cold resin as the screws rotates is largely prevented. As a result, the coefficient of friction of the barrel wall is in effect increased, resulting in improved conveying behavior and buildup of the pressure necessary for passage through the subsequent zones already in the feed zone (Fig. 6.1.8).

Cooling of this zone is necessary, however, to prevent melting of the resin under any circumstances. Fig. 6.1.9 shows the geometry of a typical grooved feed zone with tapered grooves. The high pressure buildup in the feed zone gives rise to a pressure flow in the superposition of drag flow and pressure flow shown in Fig. 6.1.5. Because of the only slight differences in velocity between the individual layers of the resulting velocity profile that is now produced, as well as the absence of a back flow, such an extruder exhibits as a rule poorer mixing than a conventional extruder. Accordingly, mixing sections usually must be employed in order to achieve satisfactory melt homogeneity.

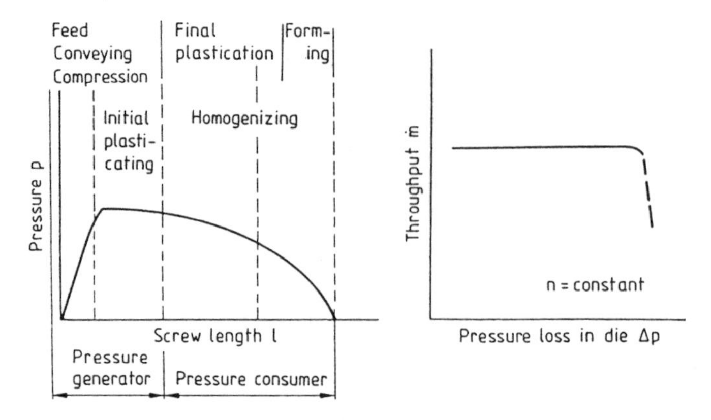

Figure 6.1.8 Pressure curve and throughput/die back-pressure behavior for an extruder with enhanced conveying (grooved feed zone).

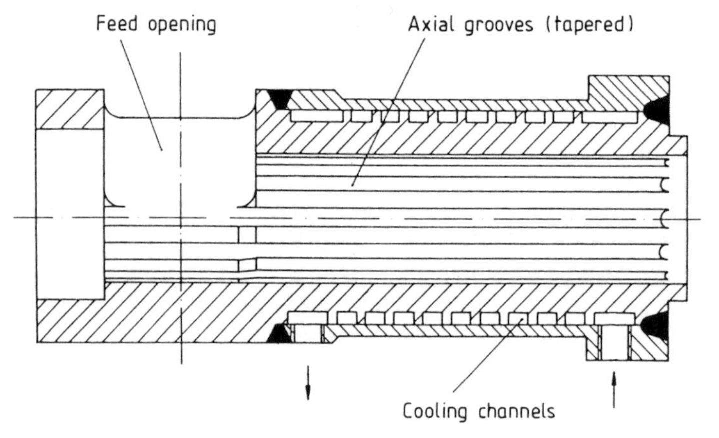

Figure 6.1.9
Grooved feed zone.

Such extruders also exhibit over wide ranges of back pressure (i.e., for different dies with different resistance values) no influence of the back pressure on the output (Fig. 6.1.8). Such extruders are said to be *forced feed extruders.*

6.1.1.3 The Plasticating Process

The heat required to melt the material is introduced primarily via two mechanisms. First, heat is dissipated as the material is conveyed, i.e., as a result of internal and external friction involving the material. Since this dissipated heat is generally not sufficient, heat is also added via the heated barrel walls. This conductive heat can be easily varied by means of electrical resistance heaters to control the melt temperature.

The manner in which the pellets melt in the screw channel can be described with the aid of models. The melting process is determined by the material behavior, with a

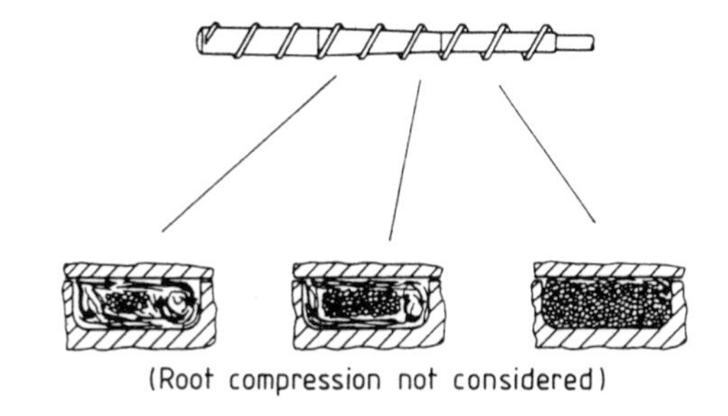

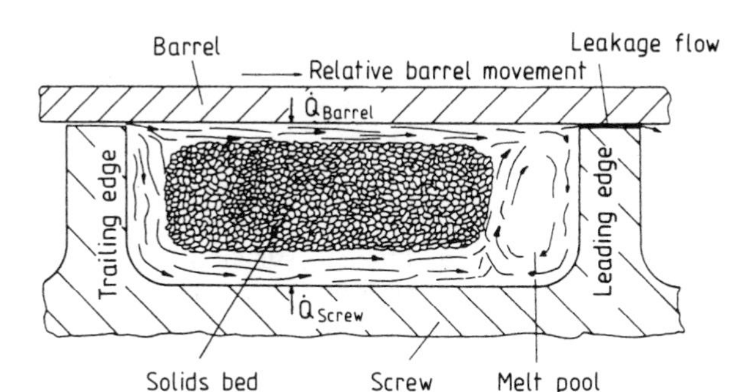

Figure 6.1.10 Melting model.

distinction being made between those materials that adhere to the wall (most thermoplastics) and those that slip (often as the result of added lubricants).

Melts that Adhere to the Wall

The material initially begins to melt on the hot barrel wall. As soon as it has reached a sufficient thickness, the resulting film is scraped off by the leading edge of the flight and collects ahead of the flight in the screw channel. The melt collecting ahead of the leading (active) edge, together with the melt introduced from the front screw channel as a result of the leakage flow, circulate around the as-yet unmelted pellets. Some of the melt is forced between the pellets as a result of the pressure created in the transition zone.

Nevertheless, a residual solids bed remains that is hardly dispersed by the flowing melt and is melted only with difficulty because of the poor heat conduction of all plastics (Fig. 6.1.10).

Melts that Slip Along the Wall

In contrast to the model presented above, the non-adhering melt film that forms on the barrel wall is forced over the screw flight in the form of leakage flow as a result of the pressure gradient. This portion of the melt is subjected to high shear and homogenized. It collects behind the trailing edge of the flight and forms a melt pool. As the result of mechanisms comparable to that for the model of melts which adhere to the wall, the remaining material is dispersed only insufficiently so that here, too, a solids bed of sintered pellets forms.

Type	Single-screw extruder	Twin-screw extruder Co-rotating	 Counter-rotating
Primary field of application	Resin compounding Plasticating (pellets)	Resin compounding	Plasticating (powder)
Conveying	Drag forces conveying	Drag forces (positive displacement)	Chamber principle
Economic advantages	Price/performance ratio Service life	Resin price Specific energy consumption	

Figure 6.1.11 Comparison of single- and twin-screw extruders.

6.1.1.4 The Twin-Screw Extruder

The family of twin-screw extruders is divided into co-rotating and counter-rotating types, where conical screws are also possible with the latter. An additional distinction results from the spatial relationship of the screw flights; they can be intermeshing or non-intermeshing. Fundamental differences with regard to the single-screw extruder are given in Fig. 6.1.11.

Co-rotating Twin-Screw Extruders

With each revolution, the co-rotating twin-screw extruder (Fig. 6.1.12) transfers the melt from the screw channel of one screw to that of the other screw. The conveying mechanism—drag forces—is comparable to that found in the single-screw extruder. By being transferred form one screw channel to another, however, the melt does follow a longer path and is subjected to higher shear.

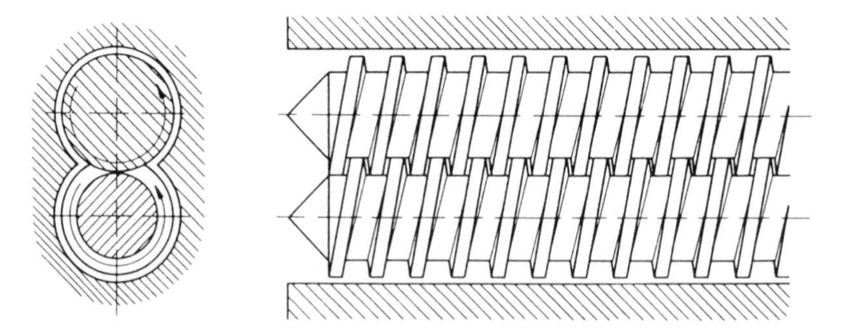

Figure 6.1.12 Tightly intermeshing corotating twin-screw extruder.

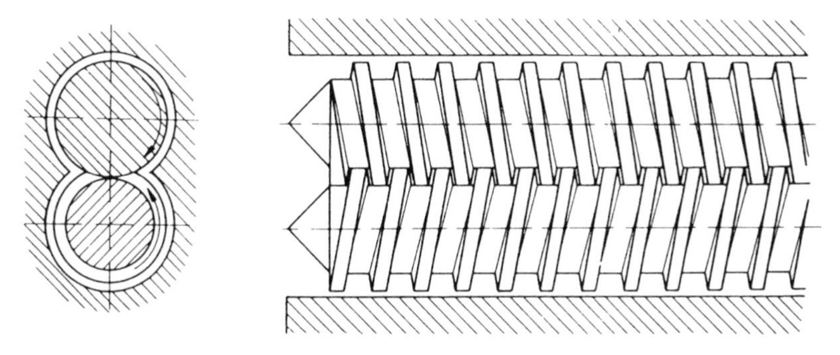

Figure 6.1.13 Tightly intermeshing counter-rotating twin-screw extruder.

Because of the design, tightly intermeshing, kneading, and shearing elements can be integrated into the screw without any problems. This type of extruder finds application today primarily in compounding.

Counter-Rotating Twin-Screw Extruders

The machines employed in profile and pipe extrusion primarily for rigid PVC are tightly intermeshing, counter-rotating twin-screw extruders (Fig. 6.1.13). A fundamentally different conveying mechanism characterizes this extruder. Each screw segment forms a closed chamber (Fig. 6.1.13) that conveys the melted material from the hopper to the end of the screw without any noteworthy exchange with neighboring chambers. Drag forces are not needed for this positive conveying, with the result that little dissipation-induced heating occurs.

This is the primary reason for its use in the above-mentioned application. Heating occurs largely via the heater bands on the barrel, which can be precisely controlled and permit gentle heating of sensitive materials. Two tightly intermeshing twin screws form C-shaped chambers between the screw flights. The C-shaped chamber is formed as the flight of one screw extends into the screw channel of the other and closes it.

As a rule, resin is metered into counter-rotating twin-screw extruders so that the screw channels (chambers) are only partially filled. Without metering, the chambers would be completely filled, resulting in increased material flow into the nip between the screws. This, in turn, leads to an increased pressure buildup, which could give rise to increased wear of the screws and barrel. The material transported via this positive conveying mechanism melts only quite slowly. It is only in the last few flights before the die that the material sinters together as a result of the built-up pressure and is forced in part through the gaps in the chambers (leakage flows). These material flows melt spontaneously due to the abrupt shear.

While this conveying and melting mechanism assures well-controlled, gentle handling of the material, the melt is usually inhomogeneous and insufficiently plasticated so that the use of mixing elements is also to be recommended here.

The twin-screw extruder is a relatively expensive machine. This is a result of the following factors, among others:

- expensive, difficult to accommodate bearings (dimensions limited because of the proximity of the screws);
- complicated gearboxes;
- barrel with figure-8 shaped bore;
- two screws.

The meshing of the screws and the high pressure built up in the final flights give rise to severe wear of screws and barrel. Machines with screw diameters up to about 160 mm are common.

6.1.1.5 Temperature Control System

The extruder is heated up and the most appropriate melt temperature for the given material is maintained by the barrel heating. For this purpose the extruder is divided into several sections (heating zones) that are heated separately in order to provide a temperature gradient along the barrel. Several temperature control systems are common today:

- electrical heating by means of resistance-type heater bands or aluminum jackets with cast-in heating elements and cooling lines or fins that are placed around the extruder barrel, sometimes in conjunction with electrically powered blowers for cooling.

and in a few cases

- heating and cooling by means of a heat-conducting medium (oil, steam, pressurized water) that is brought into direct contact with the outer wall of the barrel.

A temperature sensor is placed in the barrel wall for each heating zone and provides a temperature-proportional electrical signal that is connected to a controller. When processing thermally sensitive materials, such as PVC, the screws are temperature-controlled as well.

6.1.2 Extrusion Systems

6.1.2.1 Extrusion Die

Extrusion dies are attached to the end of the extruder and serve to form the plasticated and pressurized melt into the shape with the desired cross section. These cross-sectional shapes can be:

- solid profiles (edge moldings, rectangular bars, etc.);
- hollow profiles (shutter and window profiles, etc.);
- open profiles (channel profiles, etc.);
- pipe;
- film;

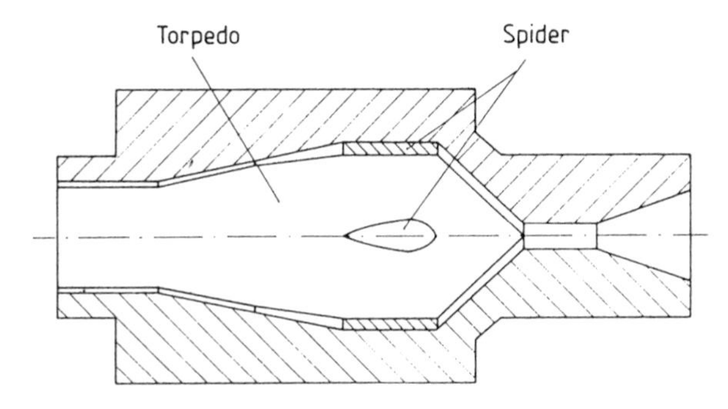

Figure 6.1.14 Spider head.

- sheet;
- filaments.

There are also dies for wire coating, for instance, as well as a number of other types.

All dies contain a flow channel through or around which the circular cross-sectional flow of melt leaving the extruder passes. The flow channel is also generally called the melt manifold. The following manifold types are of primary interest.

Spider-Type

The spider-type is the simplest melt manifold. It generally involves a streamlined-shaped torpedo (Fig. 6.1.14) connected to the outer wall of the flow channel via so-called spiders. Flow marks or weld lines in the extrudate can be caused by the melt flowing together after the spider. The torpedo is conical on the end facing the extruder and gradually changes shape as it approaches the die orifice to assume the desired cross-sectional shape.

Applications: pipe, profiles, etc.

Coat-Hanger-Type

The coat-hanger shape (Fig. 6.1.15) serves to spread the melt stream across the width. The melt first enters a manifold, which narrows toward the outside as the melt stream decreases,

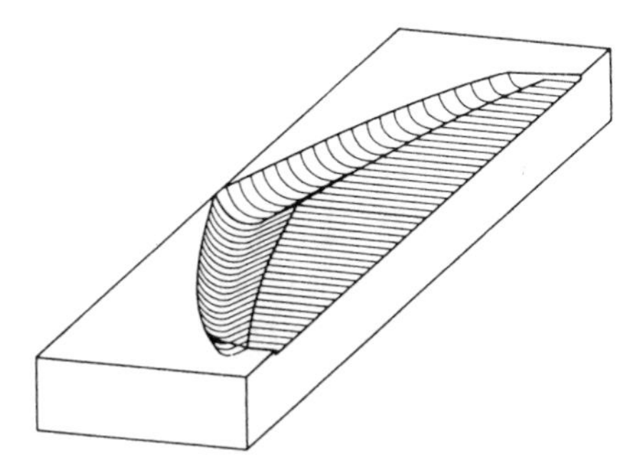

Figure 6.1.15 Coat-hanger die (manifold).

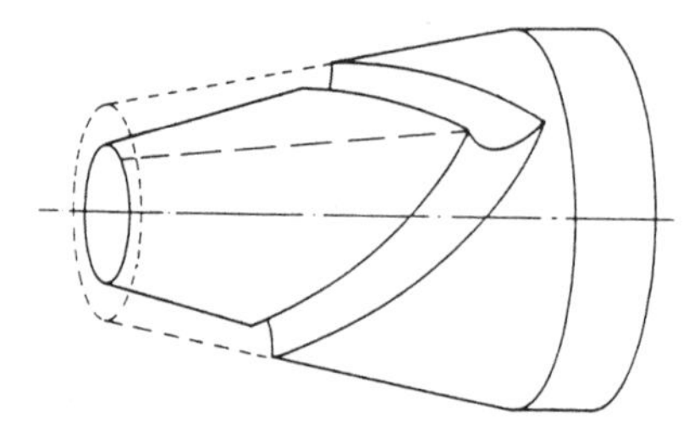

Use: Pipe extrusion, blow molding, wire sheathing

- complex manufacturing
+ design independent of operating point possible

Figure 6.1.16 Torpedo head.

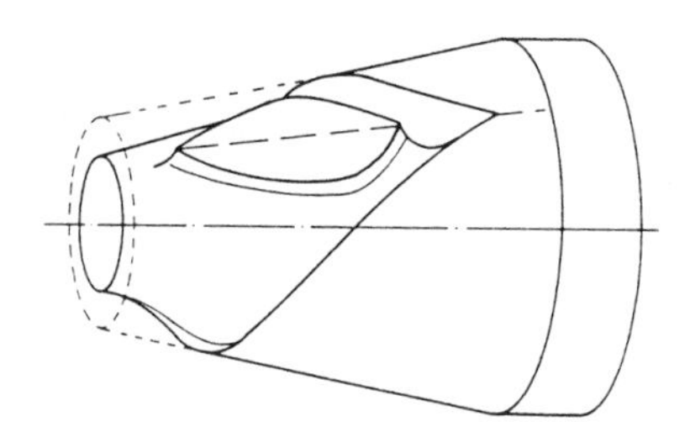

Use: Blow molding, wire sheathing

- not possible to calculate
- weld lines
- dependent on operating point
+ short design

Figure 6.1.17 Heart-shaped curve.

and then passes to the so-called island region over which it then flows. Flow control elements (barriers, adjustable lips), with the aid of which nonuniform discharge of the melt across the width can be adjusted, are located in this region. A variation of the coat-hanger manifold is the so-called fish tail manifold, in which the manifold has a geometrically simplified shape.

Applications: film, sheet, etc.

Heart-Shaped Curve (Figs. 6.1.16, 6.1.17)

The heart-shaped manifolds form a tubular melt stream. A characteristic is that melt enters from the side. A heart-shaped curve is formed by wrapping a coat-hanger manifold around a cylinder or cone.

Applications: pipe, blown film, wire coating, etc.

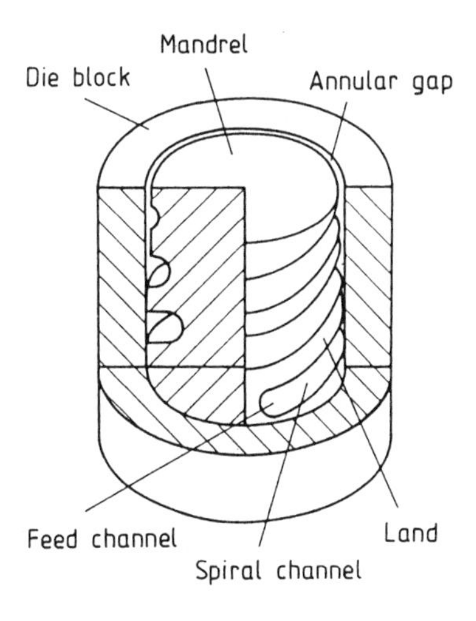

Figure 6.1.18 Spiral-type die.

Spiral-Type

The spiral-type consists of a core with several helically shaped flow channels, the depth of which decreases continuously (Fig. 6.1.18). The channels thus formed decrease in height as they approach the die orifice. The melt flow is divided into streams that flow axially and others that flow circumferentially. The end effect is an extremely homogeneous melt distributed around the circumference without flow marks and weld lines such as those that occur when the melt flows together after spiders.

Application: blown film (pipe).

As a rule, all extrusion dies are heated electrically.

6.1.2.2 Downstream Equipment

The melt leaving the extrusion die must be fixed in its shape and dimensions. This is accomplished for extrusion of profiles during the so-called calibration with the aid of compressed air or vacuum. In this step, the extrudate is pressed against the calibrator walls and cooled. After calibration, it must have solidified to such an extent that it can no longer be deformed in the following cooling zone and retains its dimensions. The length of the calibration must thus be matched to the output and geometry of the extrudate.

In contrast, flat sheet is cooled in a so-called roll stand, which also provides for calibration after leaving the die. Considerable force is required to draw the extrudate through the calibrator. The cooling in the calibrator is usually not sufficient for this force to be withstood without deformation. Accordingly, a cooling stage normally follows the calibrator. Water baths through which the extrudate passes are used for profiles, pipe, wire, and the like. Flat extrudates are cooled by rolling. Water sprays and air cooling devices are also common. Here, too, the length is matched to the output and geometry of the extrudate. The take-off, the purpose of which is to pull the extrudate at a uniform speed from the die (through the calibration and cooling), follows the cooling. Various types are available, such as belt-type, draw-roll, or caterpillar take-offs.

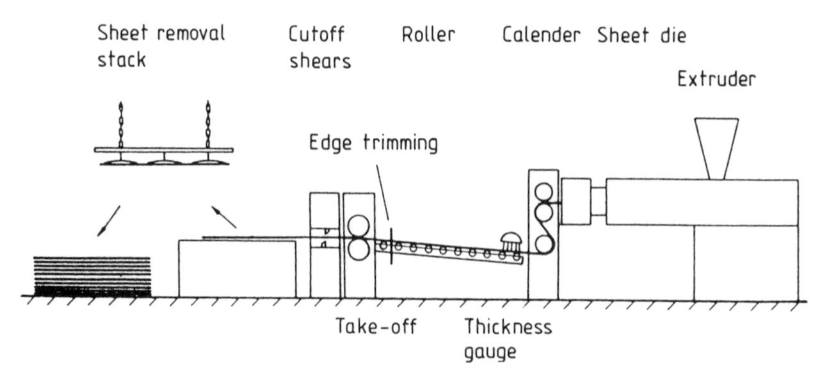

Figure 6.1.19 Sheet extrusion line (principle).

After the take-off, flexible extrudates are wound onto drums or spools (wire, filaments, film, tubing, etc.). Rigid extrudates (pipe, sheet, profiles, etc.) are cut to length with the aid of so-called cut-off devices such as saws or shears and then stacked.

6.1.2.3 Examples of Systems

Pipe Extrusion Line (Pipe Train)

The system shown in Fig. 6.1.1 is a typical example of a pipe extrusion line (pipe train). It consists of the extruder and the die in which the melt is formed into pipe. The cooling/calibration, caterpillar take-off, and cut-off device then follow.

Sheet Extrusion Line (Sheet Train)

The sheet extrusion line (sheet train) system (Fig. 6.1.19) consists of the extruder and sheet die. These are followed by the so-called calender stack. This normally contains three rolls, the surfaces and roll gaps of which serve to calibrate and cool the film (sheet). A roller train to provide air cooling and the take-off rolls are next. Finally, the sheet is cut to size and stacked.

Blown Film Line

Blown film systems (Fig. 6.1.20) are employed to produce very wide and tubular films. These systems consist of extruders with a blowing head, normally a spiral-type. The vertically extruded tubular film passes an air ring in which it is cooled. By introducing air into the interior of the tubular film, it is inflated to the desired size. At the upper end of the line it is flattened by the bubble collapsing boards while the air is retained with the aid of pinch rolls. Finally, the film is wound onto spools. A so-called calibrating basket, which defines the size or the bubble and stabilizes it, can also be employed between the air ring and the bubble collapsing boards.

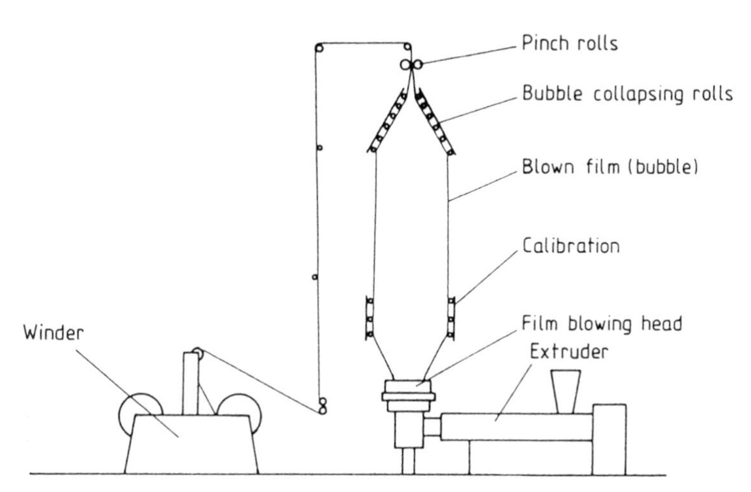

Figure 6.1.20
Blown film line.

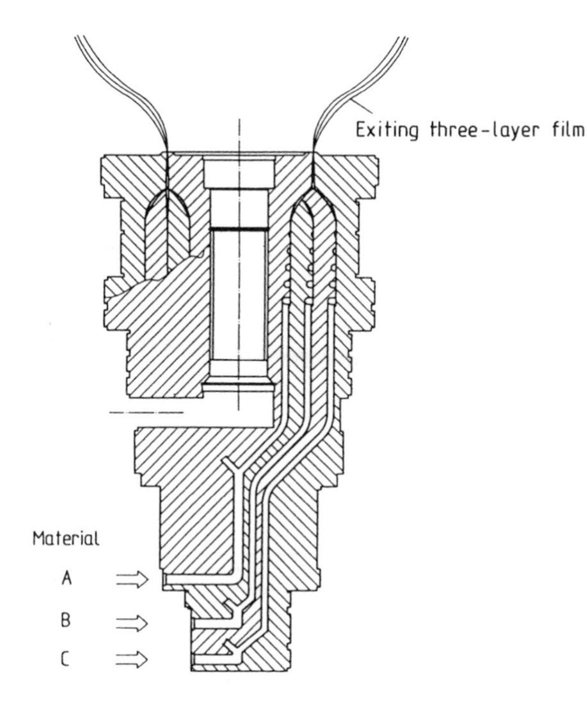

Figure 6.1.21 Three-layer coextrusion blown film die, reversing and with internal cooling (Battenfeld).

6.1.3 Coextrusion

Given the many requirements that must be satisfied by extrudates today, it is possible that these cannot be met by a single material. An attempt is made to resolve this problem by combining two or more materials in a composite structure. Multilayer wire insulating and packaging films are well-known examples.

These are produced by the coextrusion technique. Each material requires a separate extruder and melt manifold, which are mounted in succession or inside one another in the coextrusion die. Fig. 6.1.21 shows a three-layer blown film die with three nested spiral-type manifolds to distribute the melts. In the area of plastics packaging, films with up to seven layers are employed today.

6.2 Extrusion Blow Molding and Stretch Blow Molding

Hollow articles are produced in thermoplastic resins today primarily by means of extrusion blow molding and the related stretch blow molding.

The term *hollow articles* is not limited only to packaging items, such as bottles, canisters, or drums, but also includes technical parts, such as air ducts, surf boards, suitcase halves, luggage racks, or automobile gasoline tanks. The capacity of these articles ranges from a few milliliters (packaging for medication) to the present maximum of approximately 13,000 liters (heating oil tank). Fig. 6.2.1 shows the methods arranged according to features.

6.2.1 Extrusion Blow Molding

6.2.1.1 Process Sequence

Production by means of extrusion blow molding consists of two processes occurring in parallel:

- the continuous extrusion of the parison (initial step),
- the cyclic transfer of the parison and forming in the mold with the aid of the blowing air (second step).

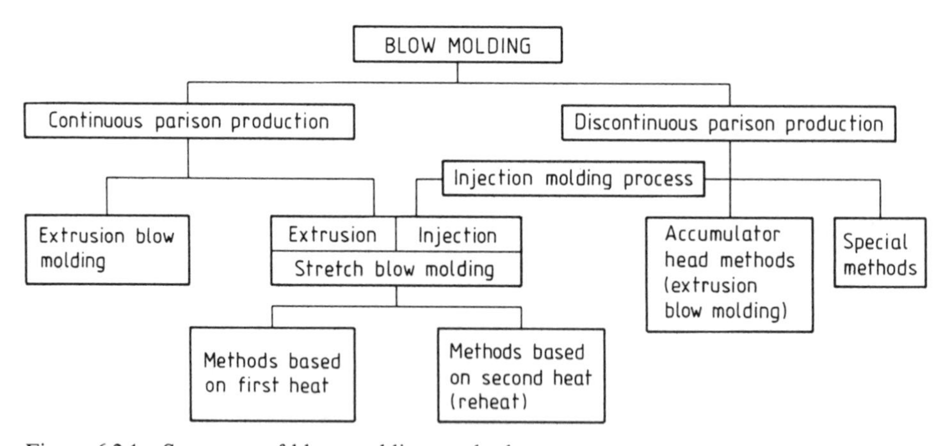

Figure 6.2.1 Summary of blow molding methods.

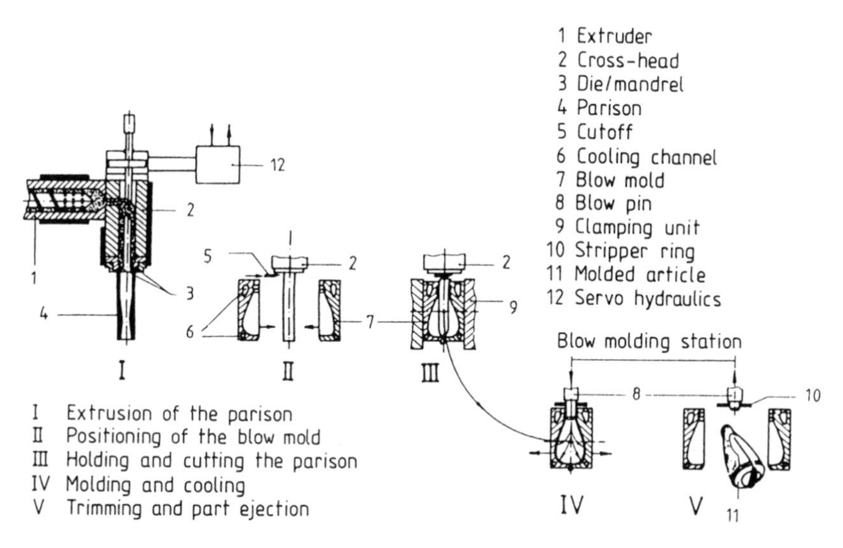

Figure 6.2.2 Process sequence for extrusion blow molding.

Fig. 6.2.2 shows the entire process broken down into individual steps. The first step begins when the extruded parison has the necessary length. The mold then closes around it and a knife cuts the parison. Next, the mold moves to the blowing station. Here the blow pin is introduced, and the actual blow molding takes place with the aid of the compressed air that flows through the blow pin. After the cooling time has elapsed, the mold opens and the finished article is ejected. The mold then returns to its position beneath the extrusion head in order to receive a new parison. With very large and heavy molds, the shuttling step is eliminated by having the parison cut off and transported to the mold by a gripper.

Part ejection and trimming of the flash usually take place automatically. The flash represents material caught between the pinch-off edges as the mold closes. Depending on the shape of the molded part, the flash can vary from a few percent up to several times (in extreme cases) the molded part weight. For this reason, an attempt is made to feed this material into a granulator immediately so that it can be returned to the extruder as so-called regrind.

The polyolefins (HDPE, LDPE, PPE) are the most commonly processed materials. At present, a large amount of PVC is also processed, but this is declining.

6.2.1.2 *The Machine*

An extrusion blow molding machine consists primarily of five major components (Fig. 6.2.3):

- machine frame with clamping unit and blowing station,
- mold,
- extruder,
- extrusion head,
- control cabinet.

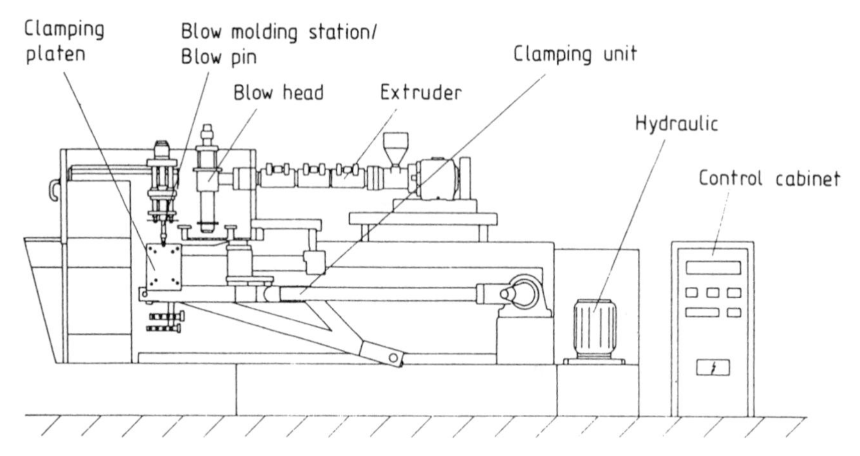

Figure 6.2.3 Extrusion blow molding machine.

The machine frame provides the basic framework for the machine and always contains as a component the clamping unit. Its functions include: opening and closing the mold halves (similar to an injection molding machine) as well as transporting the mold between the extrusion head and the blowing station. In comparison to an injection molding machine, the clamping unit is generally of a lighter design, since the molding clamping forces are relatively low as a result of the lower internal mold pressures.

The clamping unit is driven by the hydraulics, which are also located in the machine frame. The blowing station describes that portion of the machine in which the mold stays when the blow pin is introduced and the actual blow molding process takes place.

The function of the blow pin is to introduce the blow air while forming and calibrating the neck region of the article. Fig. 6.2.4 shows a closed mold with a parison (preformed) inside and the blow pin in position. The latter is normally hydraulically actuated and water-cooled.

The Mold

The mold generally consists of two halves into which the negative contour of the molded part has been machined. Its function includes shaping and cooling the molded part and forming the pinch-offs. The latter arise since the parison is a section of an extruded tube that naturally cannot be blow molded until its ends are sealed to prevent the blowing air from escaping. In the example shown in Fig. 6.2.4, this is accomplished at one end by the blow pin. At the other end, the parison is pressed closed during the closing motion by the pinch-off edges machined into the parting line of the mold, thereby welding the molten material. These pinch-offs are a characteristic feature identifying extrusion blow molded articles. Thus, a prerequisite of the materials employed is that they weld easily. The mold can cool the molded part only from one side. To shorten the cooling times, the blow pin contains an additional orifice through which the blowing air exits via a restriction—thereby

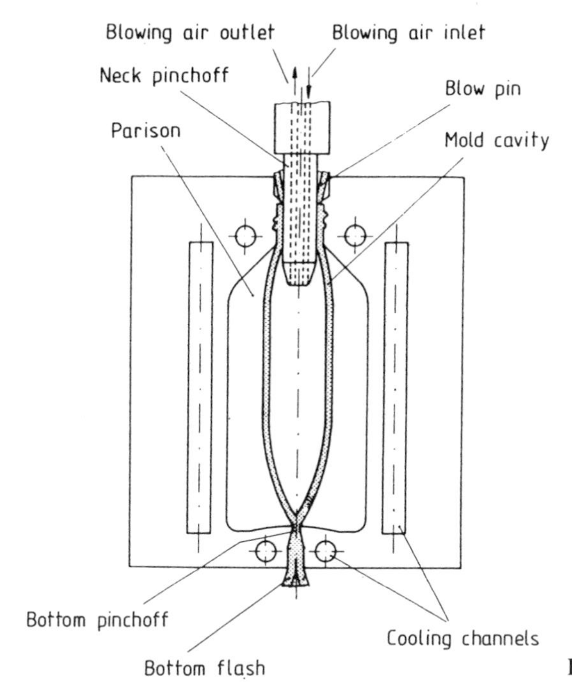

Figure 6.2.4 Blow mold.

building up pressure and creating a circulatory flow of air in the molded part that provides a certain amount of heat removal (flushing air method).

So-called intensive cooling methods also provide a reduction in cooling time. For these methods, either cryogenic air is employed as the blowing medium or liquefied gases (nitrogen or carbon dioxide) are added to the blowing air.

Since the blowing pressures seldom exceed 4–8 bar (60–120 psi), the mechanical design of the mold does not have to satisfy very high requirements. Because of its good thermal conductivity, aluminum is a preferred material.

The Extruder

As with other processing methods, the function of the extruder is to provide a thermally homogeneous melt at the desired discharge rate and pressure. Blow molding machines employ almost exclusively single-screw extruders with a screw length of 20–25 D. Depending on the material to be molded, a grooved and cooled feed zone may be a standard feature.

The Blow Head

The so-called blow head is attached to the end of the extruder. Its functions include changing the melt flow from horizontal to vertical and forming the parison. Centrally fed spider-type manifolds (strainers), annular channels, heart-shaped curves, torpedoes, and spiral-type manifolds distribute the melt (see the chapter on “Extrusion”).

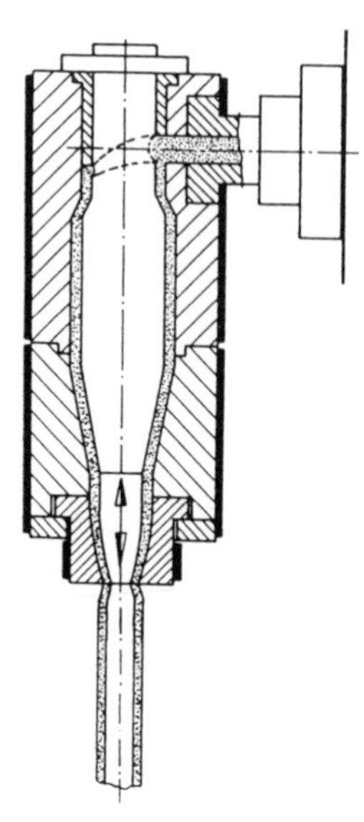

Figure 6.2.5 Torpedo head.

The flow channel at the discharge is generally conical in shape, with either the inner or outer conical component movable. As an example, Fig. 6.2.5 shows a torpedo head in which the inner core can be moved axially with the aid of servo hydraulics. The objective of this motion is to continually change the wall thickness of the parison during extrusion. The purpose of this device, which is used for parison programming, can be seen in Fig. 6.2.6.

If the bottle sketched were to be blow molded from a parison with a constant wall thickness, the blow molded bottle would have a nonuniform wall thickness distribution because of the different amounts of stretching. This has two serious disadvantages.

First, unnecessary material is consumed, since the wall thickness must be sized in accordance with the thinnest location. If one considers that, for mass-produced articles such as beverage bottles, approximately 50–60%—up to almost 70% when using very expensive materials

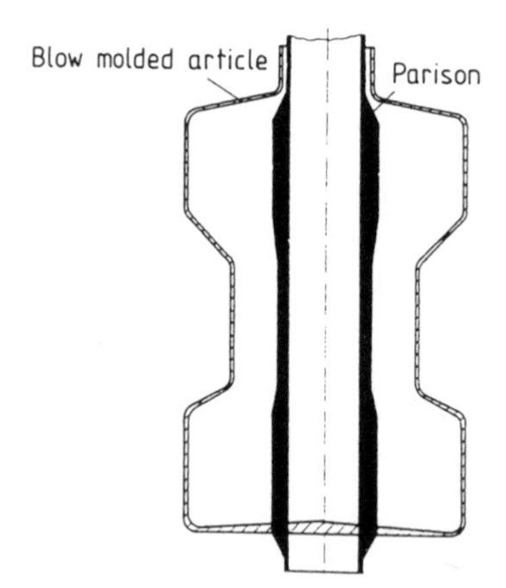

Figure 6.2.6 Parison with profiled wall thickness.

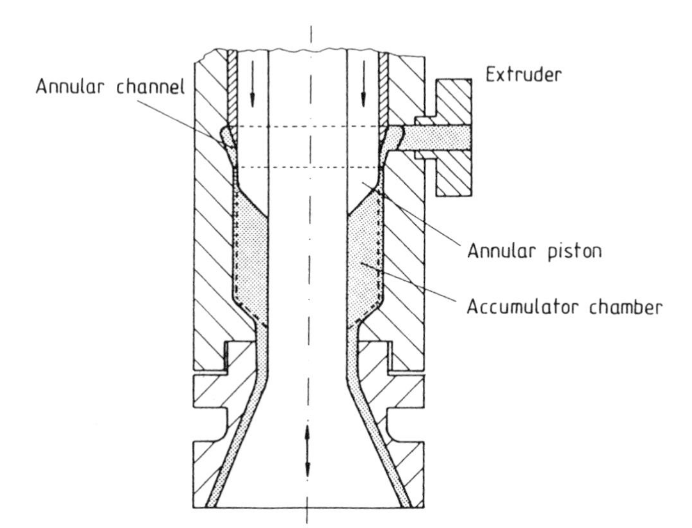

Figure 6.2.7
Accumulator head.

(e.g., polycarbonate)—of the manufacturing costs are attributable to the material, this represents an effective means for reducing costs.

The other disadvantage is that unnecessary thick sections require correspondingly long cooling times (the cooling time is proportional to the square of the wall thickness), since even the section with the thickest wall must be cooled down to at least the maximum demolding temperature. The consequences are unnecessarily long cooling times in conjunction with correspondingly high production costs. A distinction is made between continuously operating blow heads and so-called accumulator heads, the use of which is necessary in the following applications. For large parts, the preform must have a corresponding circumference and length. This leads to the problem that the extruded molten parison stretches under its own weight—the wall thickness is uncontrollable—and under certain circumstances can even break. In addition, the lower end is already considerably cooler than the upper end—the behavior of the material during blow molding is length-dependent. In short, the time to produce the preforms/parison is too long. A remedy to this is provided by so-called accumulator heads, which are filled by one or more continuously operating extruders. If the mold is beneath the head ready to accept the parison, the melt is discharged from the accumulator within the shortest period of time so that the above-mentioned problems can no longer occur. Fig. 6.2.7 shows one of the various types of accumulator heads in which the melt is collected in an annular chamber and then discharged hydraulically by means of an annular piston.

Common to all types is that the continuously supplied melt is accumulated and the parison is formed by cyclic emptying of the accumulator. The use of parison programming is also possible with accumulator heads.

6.2.1.3 Multiple-Cavity Blow Molding

To increase the output of blow molding machines, particularly for smaller articles, two approaches are taken. On the one hand, multiple heads are employed in which the melt stream coming from the extruder is divided into a maximum of four secondary streams that simultaneously produce as many parisons. These are then placed in multicavity molds, i.e., several articles are produced during one cycle.

The second approach uses several molds in succession that accept the parison, mold the part, and cool it. A common design is the so-called *tandem machine*, with two clamping units, in which one mold is beneath the extruder head and accepts a parison while the other mold is under the blow pin.

Nothing changes in the basic process sequence for either type of machine.

6.2.1.4 Coextrusion

The various requirements that must be met by packaging today often cannot be fulfilled by a single material. For this reason, the technique of coextrusion, where different materials are fed from the corresponding extruders into the extruder head, flow together there, and are discharged as a multilayer parison, is employed in such cases. This parison is then molded quite conventionally. There are in operation today coextrusion blow molding systems in which the parison consists of up to seven layers. Besides the primary materials, resins such as polyamide/nylon (PA), polycarbonate (PC), PET, PVDC, EVOH, and others are used. Since these materials generally do not adhere to one another, a so- called adhesive (bonding) layer must be extruded between them.

6.2.2 Stretch Blow Molding

Stretch blow molding is a special variation of blow molding. It utilizes the effect— similar to that in films—that by stretching near the glass transition or crystallite melting temperatures a high degree of orientation can be introduced into the material. In this way the mechanical properties are improved considerably. In this process, the parison is stretched not only radially (as in extrusion blow molding) but also longitudinally. The latter is accomplished mechanically with the aid of a core, while the circumferential stretching is accomplished simultaneously with the blowing aid. The relatively low temperatures at which this process takes place require high molding forces, for which reason blowing pressures of up to 20 bar (290 psi) are encountered.

In actual practice, it is generally necessary to find a compromise between the ideal stretching temperature and attainable stretching forces. The basic sequence of the stretch blow molding process is shown Fig. 6.2.8, starting with an injection molded preform. PVC, PP, and PET are the preferred materials for this process.

6.2.2.1 Production of Preforms

Preforms of PVC, for example, that are difficult to produce by means of injection molding are produced by means of extrusion blow molding and then processed on stretch blow molding machines. Injection molded preforms have the advantages of a very accurate wall thickness distribution and a very precisely calibrated neck. In addition, there are no pinch-offs; such preforms can, however, be recognized by the gate vestige.

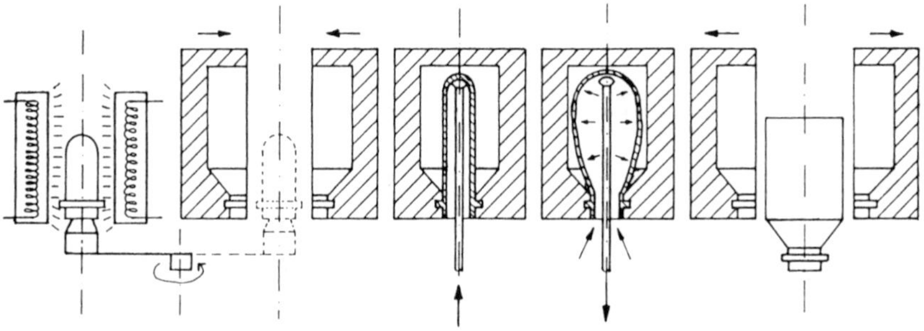

Figure 6.2.8 Principle of the stretch blow molding process.

6.2.2.2 *Stretch Blow Molding in the First and Second Heat*

The following process sequence describes stretch blow molding in the first heat:

Production of preform—cooling to the stretching temperature (conditioning)—stretching

This takes place in one machine (injection stretch blow molding, extrusion stretch blow molding machine). The second version is stretch blow molding in the second heat, which takes place on two separate machines in the following sequence:

Production of preform—cooling to demolding temperature—reheating to stretching temperature—stretching

The first method involves less energy, but the machinery is more complicated. The difficulty with both methods is achieving a constant stretching temperature across the wall thickness of the parison. The wall thicknesses of stretch blow molded preforms can be up to several millimeters, since in order to achieve longitudinal stretching the preform must be several times shorter than the finished product.

6.2.2.3 *Advantages and Disadvantages of Stretch Blow Molding*

A hollow article produced by means of stretch blow molding is characterized by the biaxial orientation of its macromolecules and the resulting property improvements. These include above all the higher strength values versus a conventional blow molded article with identical wall thickness. Depending on the application and material, there may also be improved transparency, higher surface gloss, and better barrier properties with respect to various substances.

The restrictions with regard to part shape are a disadvantage. In general, only rotationally symmetrical parts or those with an oval cross section can be produced.

6.3 Injection Molding

Injection molding represents one of the most important production methods in the processing of plastics. Injection molding has several features that make utilization of this method advantageous, especially for the mass production of complicated parts. These features are

- direct path from resin to finished part,
- no or only minimal finishing of the molded part necessary,
- process may be fully automated,
- good reproducibility of production.

The range of parts that can be produced by injection molding extends from the smallest gears and bearings to large waste containers. The weights of molded parts are on the order of 10^{-6}–10^{2} kg.

6.3.1 Machine and Process Sequence

According to German standard DIN 24 450, the function of an injection molding machine is the discontinuous production of molded parts, preferably from macromolecular molding compounds, with the forming taking place under pressure.

Process Sequence

To accomplish the above, it is necessary to convert the material from the as-delivered condition—usually in pellet or powder form (elastomers are generally supplied in strip form)—into a melt-processible state.

In the first step of the process, the material is plasticated by the rotation of the screw (Fig. 6.3.1). After closing of the mold (step 2), which contains a cavity in the shape of the part to be molded, the plasticated material is injected through axial displacement of the screw (step 3).

With thermoplastic materials, the melt is subsequently cooled in the mold (step 4). In the case of cross-linking materials (thermosets, elastomers), the mold is heated and thus initiates the cross-linking reaction.

The last process step involves opening the mold and ejecting the molded part (step 5).

The process steps sketched, which also overlap somewhat in time (Fig. 6.3.2), are a part of each injection molding cycle. During production of an injection molded part, the individual production steps, which are shown once again in greater detail in Fig. 6.3.2, are coordinated by the machine controls. For economical production of a molded part, the objective is to reduce the cycle time necessary for this sequence as much as possible in order to achieve high production output.

In addition to the above-mentioned process steps, other functions may take place, such as retraction of the injection unit, actuation of cores, among others, but these do not change the basic sequence.

1. Mold closing

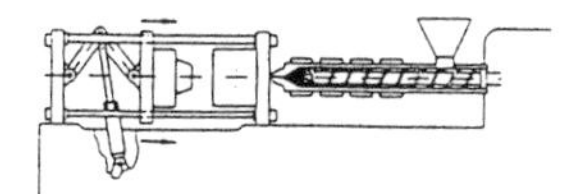

2. Injection

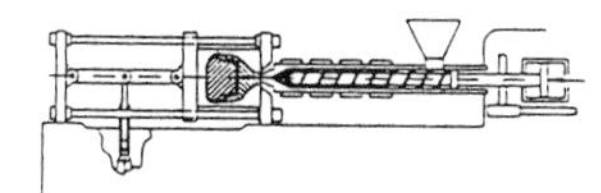

3. Holding pressure and cooling phase

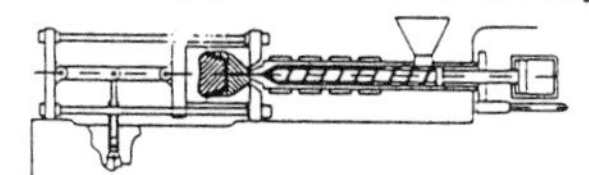

4. Recovery

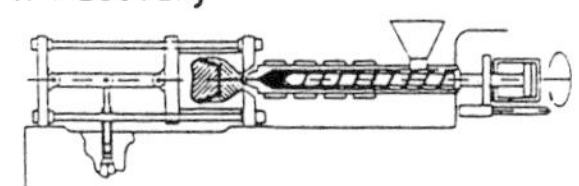

5. Mold opening and part ejection

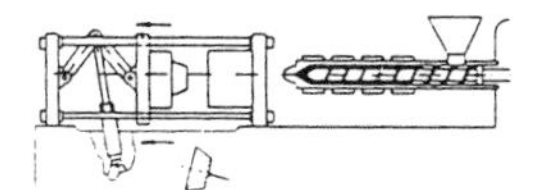

Figure 6.3.1 Process sequence for injection molding.

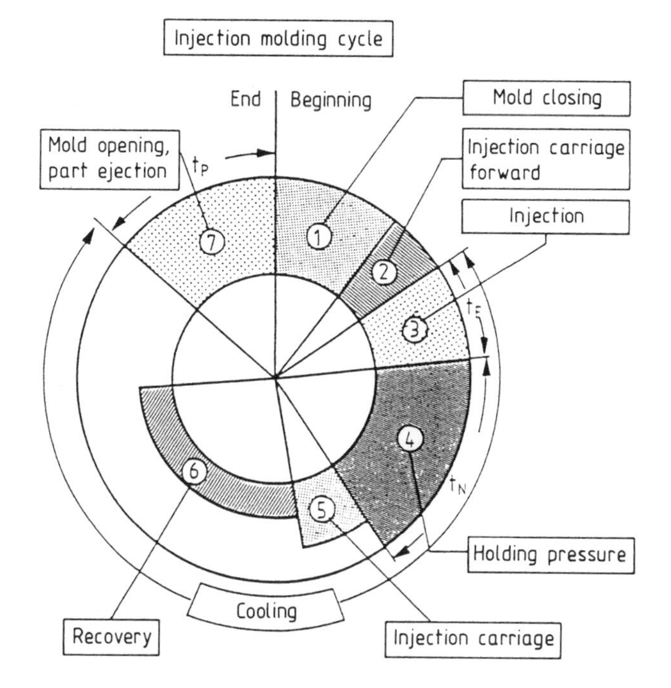

Figure 6.3.2 Injection molding cycle.

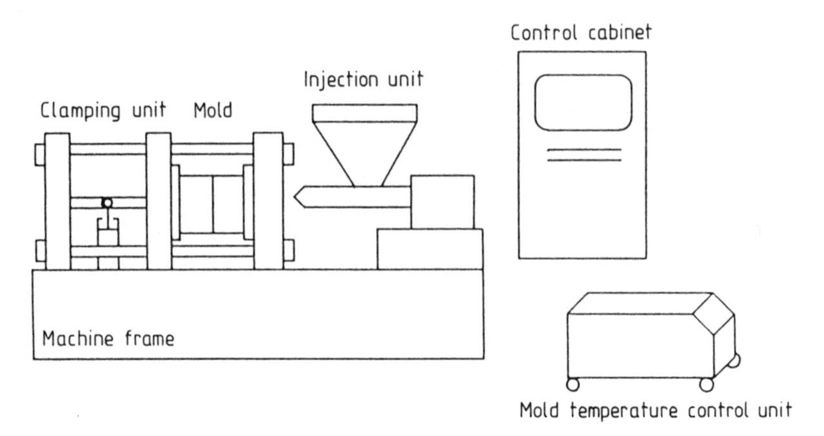

Figure 6.3.3 Components of an injection molding machine.

Machine Components

Regardless of the material to be processed, an injection molding machine consists of the following components (Fig. 6.3.3). The *machine frame* supports the *injection unit* and the *clamping unit*. The latter serves to open and close the mold during the production cycle. The sequence of the injection molding cycle is coordinated by a *control system*, which is usually located in a control cabinet separate from the machine.

As a rule, the clamping unit as well as the screw in the injection unit are driven hydraulically. The pumps required to provide the necessary flow of hydraulic oil are located in the machine frame and are powered electrically. The hydraulic valves needed to control the individual motions are located if at all possible in the immediate vicinity of the associated actuator.

To actuate additional functions, such as ejectors, shutoff nozzles, or core pulls, purely electromechanical or pneumatic drives are employed additionally. However, the following discussion will focus only on machines whose primary mechanisms of motion (plasticating, injection, mold opening/closing) are accomplished hydraulically, since this is the type of machine most commonly found in actual use.

To cool the molten injected plastic, the molding compound must be cooled to below the solidification point, if it is a thermoplastic resin, before the molded part can be removed. (With cross-linking molding compounds, the mold is heated in order to trigger the cross-linking reaction.) For this, mold temperature control devices (also called mold temperature controllers) are necessary. These can be either a fixed part of the machine or a separate unit external to the machine.

Machine Data

A variety of data can be specified to classify the machine size and performance capability. For an initial rough characterization, however, the maximum *clamping force* F_s is usually given. This is the maximum force with which the two mold halves are pressed together in order to prevent opening of the mold as a result of the internal pressure during the filling and holding pressure phases. Often, the diameter of the plasticating screw d_s serves as a characterizing feature.

In addition, the maximum possible *injection pressure* p_{Emax} is of interest. Since the diameter of the screw for a given machine has a direct effect on the maximum achievable injection pressure and the injection capacity, the product of the maximum injection capacity V_{Dmax} and the maximum injection pressure p_{Emax} is given as a characteristic value.

The injection pressure, clamping force, and screw diameter of commercially available machines lie in the following ranges:

P_{max}	1500–2500 bar (21,000–36,000 psi)
F_s	2×10^2–10^5 kN (22–11,000 tons)
d_s	20–120 mm (0.75–5 in)

The size designation of a machine usually contains information on the clamping force F_s and the product P:

$$P = \frac{V_{Dmax}\,[\mathrm{cm}^3] \cdot p_{Emax}\,[\mathrm{bar}]}{1000}$$

Thus, a machine with the designation

"Company XYZ 150/500"

represents a model from the XYZ company with a maximum clamping force of 150 tons and a value of $P = 500\ \mathrm{cm}^3$ bar.

The dimensions of the machine also increase in relation to the clamping force. By order of magnitude, these range

	Length	Width	Height	
from	100	25	60	cm
to	2500	500	500	cm

and more.

6.3.2 Components

In the following, the function of the individual components and their place in the molding cycle will be discussed for injection molding of thermoplastics. Deviations associated with the processing of thermosets and elastomers are discussed at the appropriate time.

6.3.2.1 Injection Unit

With regard to the molding compound, the functions of the injection unit of an injection molding machine (Fig. 6.3.4) are to

a) plasticate,

b) convey,

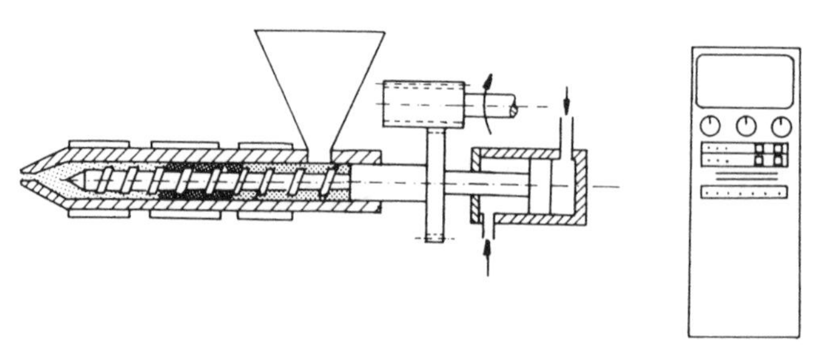

Figure 6.3.4 Conventional injection unit.

c) homogenize,

d) meter (recover),

e) accumulate,

f) inject.

In order to satisfy these requirements as fully as possible, so-called general-purpose screws are usually employed in injection molding. These have the following three zones:

- feed zone,
- transition (compression) zone,
- metering zone.

In addition to these, thermoset and elastomer screws, preplasticating screws, and others designed for specific purposes find application. The specifics of the various types are discussed in general in Section 6.1.

Recovery

Just as in an extruder, the material in the barrel of the injection unit is conveyed by the rotating screw from the hopper to the screw tip.

Thermoplastics are melted primarily by the *frictional heat* that arises during rotation of the screw as a result of the friction between the material and the screw or barrel wall. The heater bands attached to the outside of the barrel provide an additional amount of the necessary heat through conduction. Since thermal energy can be introduced by means of conduction during portions of the cycle when the screw is idle, the screws used for injection molding are usually shorter than those in extruders (20 D). This segment of the cycle, identified as the *recovery phase* in Fig. 6.3.2, must logically have taken place at least once before the first cycle begins, in order that the amount of molding compound necessary to fill the first part be available. During recovery, the nozzle orifice at the end of the injection unit facing the mold is closed off either by the sprue attached to the molded part still in the mold (recovery with injection unit in the forward position; open nozzle) or by special devices called *shutoff nozzles* (Fig. 6.3.5). Since the conveyed material requires space, the screw is displaced backwards by the accumulating melt. The space in front of the screw—also called the *melt chamber*—fills with melt.

Melt is conveyed until the screw reaches a position specified numerically in the control unit or defined by a limit switch. The distance moved by the screw to reach this point is called

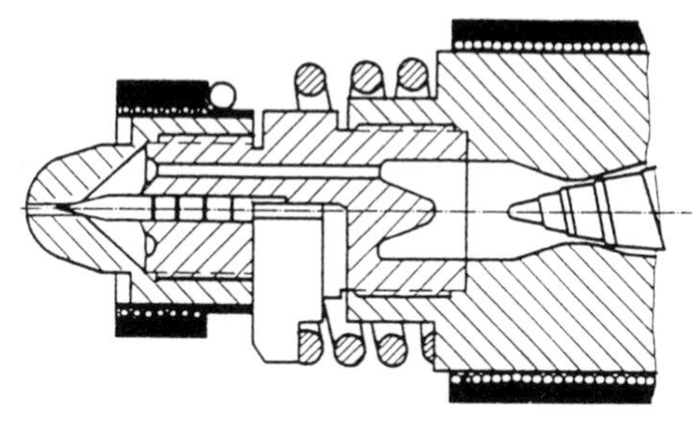

a) Needle shutoff nozzle (spring loaded)

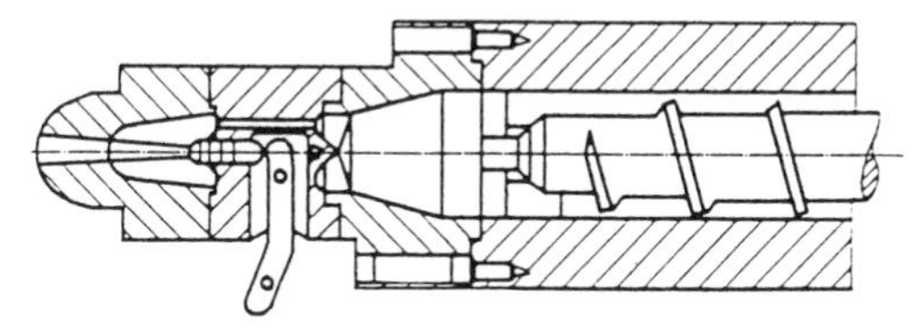

b) Needle shutoff nozzle (externally actuated)

Figure 6.3.5 Shutoff nozzles.

the *recovery stroke*, and the volume of material conveyed during the recovery phase is called the *shot volume*. The recovery stroke and shot volume differ from mold to mold and must be carefully adjusted when setting up the machine. As a rule, a slight pressure, the so-called *back pressure*, is applied during the recovery phase to the hydraulic piston that advances the screw axially during injection. Because the screw must work against this back pressure as it retracts, better melt *homogeneity* is achieved. With most materials, some minimum value of back pressure is required during recovery to avoid entrapped air.

The back pressure must also be set at the machine in accordance with the specific material processed. The pressure that results in the melt chamber because of this measure can be up to 400 bar (5800 psi). Excessive back pressure, however, can result in opening of the shutoff nozzle during recovery, resulting in material leakage. This must always be prevented.

If recovery takes place with the injection unit in the forward position up against the sprue, the pressurized melt will shoot into the opened mold as the part is ejected upon conclusion of recovery. To prevent this, the pressure in the melt chamber must be reduced. This is accomplished by pulling back the screw axially by a few millimeters after recovery. This retraction stroke is called the *decompression stroke*.

When a needle shutoff nozzle is used, the injection unit can be retracted from the mold as soon as the holding pressure phase ends (Fig. 6.3.6). Recovery then takes place with the injection unit in the "retracted" position. When an open nozzle is used, the injection unit can be retracted only after the decompression stroke. It goes without saying that, prior to injection, the injection unit must once again be brought forward against the mold.

The primary purpose for retracting the injection unit from the mold is to avoid undesirable heat flow from the injection unit to the mold, or vice versa. As already mentioned, the

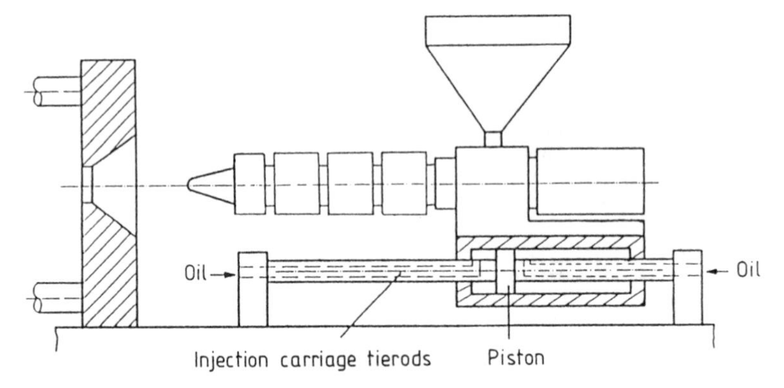

Figure 6.3.6 Injection unit supported on tierods with integrated traversing piston.

nozzles employed are either *open nozzles* or *shutoff nozzles*. The latter can be designed as follows:

- needle shutoff nozzles,
- sliding shutoff nozzles,
- cross pin shutoff nozzles.

Actuation is accomplished either mechanically or hydraulically. For a better understanding, the operation of the above-mentioned nozzles (shown in Fig. 6.3.5) will be discussed briefly.

The *needle shutoff nozzle* can be designed either as a spring-loaded nozzle (Fig. 6.3.5a) or as an externally actuated nozzle (Fig. 6.3.5b). In both cases, a needle inside the nozzle is used to close the nozzle orifice. The melt flows from the melt chamber to the nozzle tip via appropriate channels. In the spring-loaded version, the needle is pushed back against the spring force by the pressure that builds up during injection, thereby opening the nozzle orifice.

This means, however, that there is an additional pressure loss in the nozzle, resulting in increased dissipative heating of the melt. To avoid this, the externally actuated (usually hydraulically) shutoff nozzle is employed. Here the needle is opened and closed from the outside via a linkage, thus reducing the dissipation. A disadvantage of this type of nozzle, however, is the usually greater length.

The *sliding (floating) shutoff nozzle* (Fig. 6.3.7b) is opened by the nozzle contact pressure as the injection unit is brought forward against the mold. As the injection unit retracts, spring force causes the front section of the nozzle to slide, closing off the flow channel. Compared to the needle shutoff nozzle, the pressure losses are lower.

In addition to the possibility of closing the nozzle by means of axial displacement of a shutoff mechanism, this can also be accomplished by sliding a pin radially. This principle is employed in the *cross pin shutoff nozzle* (Fig. 6.3.7a). Compared to the other types, however, increased space requirements and costs are associated with this design. When using shutoff nozzles, it must always be kept in mind that these contain "dead corners" in which material can collect and remain for a very long time. With thermally sensitive materials, molecular degradation can take place after excessive residence times. If particles of the degraded

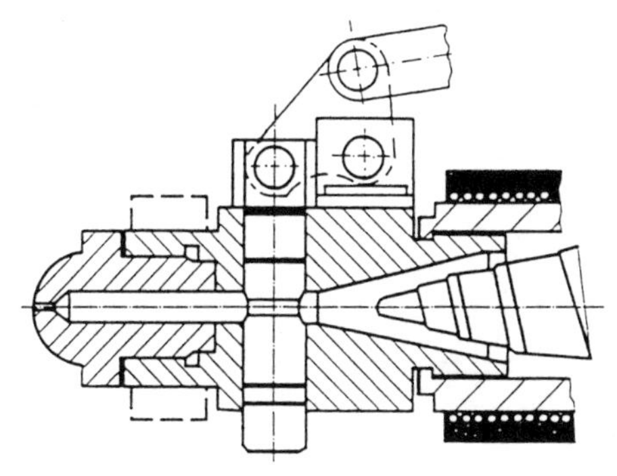

a) Cross pin shutoff nozzle

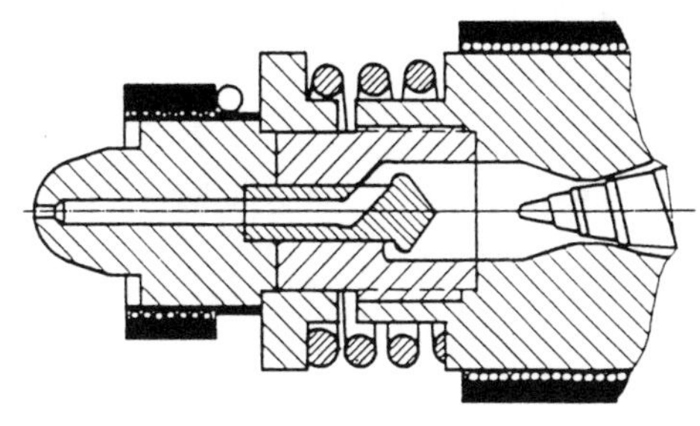

b) Sliding shutoff nozzle

Figure 6.3.7 Shutoff nozzles.

material are dislodged from the corner in course of time, black marks in the molded part can result. Shutoff nozzles can be employed to only a limited extent with these materials. When processing thermosets or elastomers, they should not be used because of the possibility of initiating cross-linking.

Injection

During injection, the screw is advanced without rotation (!) by the hydraulic injection cylinder, thereby acting as a piston to force the accumulated melt through the nozzle into the mold. The *screw advance speed* is controlled in this phase by the flow of oil provided by the hydraulic pump. It is possible to specify a profile for the speed, which is then followed by the machine controls. The hydraulic pressure in the injection cylinder is, as a rule, not constant during the *injection phase*, but it increases during the course of injection. The reason for this is that, to maintain the flow process in the mold with increased filling, higher pressure is necessary to provide a constant injection rate. During this segment of the molding cycle, operation of the machine is velocity-controlled. As a consequence, the hydraulic pressure increases. This situation often leads to difficulty with regard to understanding certain machine setup parameters, e.g., the value of the *injection pressure*. This is set at the machine as a fixed value and refers either to the hydraulic pressure p_H or the pressure in the melt

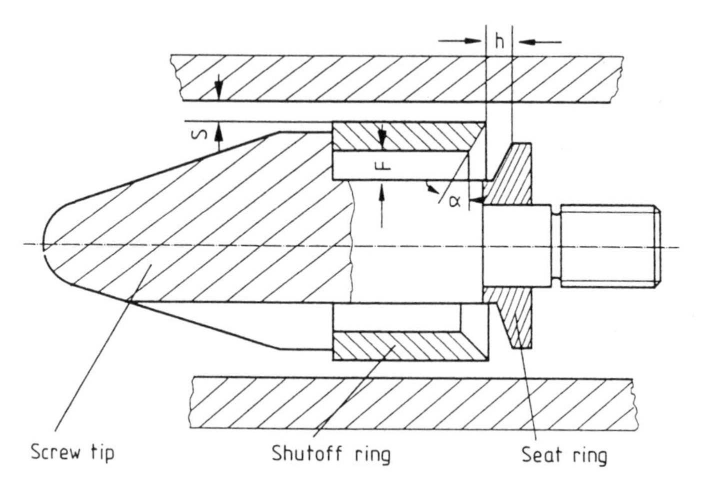

Figure 6.3.8 Nonreturn valve with sliding shutoff ring.

chamber p_s, which is obtained from the ratio of the area of the injection cylinder A_H to the area of the screw A_s:

$$p_s = \frac{A_H}{A_S} \cdot p_H$$

As mentioned, however, p_s is not constant.

In reality, the set injection pressure represents an *upper limit* that cannot be exceeded during injection. If the hydraulic pressure reaches this limit prior to completion of injection, the capacity of the hydraulic system is insufficient to increase the pressure further. The consequence is that the screw advance speed drops below the specified set point. The pressure that can be achieved in the melt chamber of commercially available machines lies between 1500 and 2500 bar (21,000 and 36,000 psi), and even higher in special machines.

Since during injection there is no screw rotation, which would mean an increase in pressure toward the screw tip, there is the danger that some of the accumulated melt will flow back over the screw flights when pressures are high in the melt chamber. This would make reproducible operation of the machine difficult and could lead to undesirably long residence times for the resin in the hot regions of the injection unit. A common remedy for this in injection molding is the use of a *nonreturn valve* on the screw tip (Fig. 6.3.8). The most common design employs a moving check ring that is forced against a seat ring at the beginning of injection by the forward motion of the screw. This closes the flow channels through which the melt passes as it moves from the screw channel into the melt chamber during recovery.

Back flow is thus prevented. It is only through the clearance between the ring and barrel wall that slight leakage flows can occur, which, however, can be taken into account.

Holding Pressure Phase

After the part has been filled volumetrically, compression of the melt takes place (*compression phase*). The pressure in the cavity increases rapidly (Fig. 6.3.9). In the melt chamber, it can

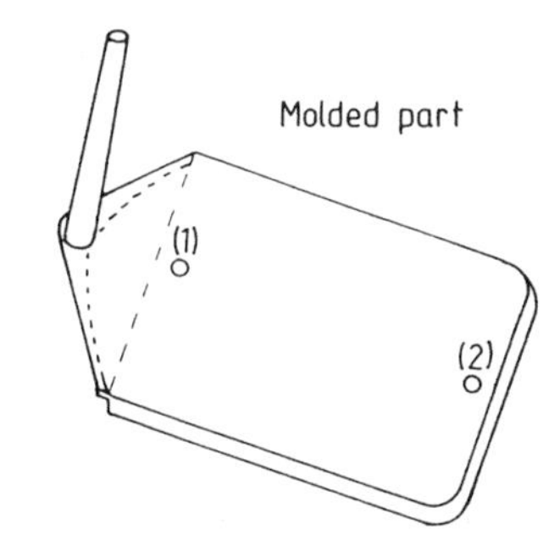

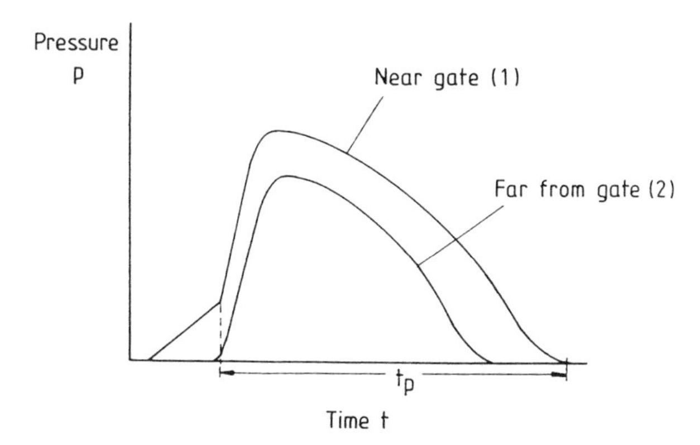

Figure 6.3.9 Pressure curves measured near gate and far from gate in a plaque mold.

even reach the set injection pressure, which usually leads to impermissibly high pressures in the mold. Accordingly, there is a transfer from the velocity-controlled to the pressure-controlled phase, the *holding pressure phase*, upon reaching a specified pressure level in the mold.

The pressure in the hydraulic cylinder is controlled in accordance with a specified set point, which can either be a constant value or profiled. The screw advance speed during this phase is quite low and results from the need for a slight flow of melt into the molded part to compensate for shrinkage during cooling.

The *transfer* from injection to holding pressure can be accomplished in several manners. The possibilities are basically:

- time-dependent,
- screw position-dependent,
- hydraulic pressure-dependent,
- cavity pressure-dependent.

The best, but also the most involved, is the latter, since this provides the greatest reproducibility. Time-dependent transfer should be used only as an exception, since this can easily lead to overpacking of the mold. Machines today provide only position- and/or pressure-dependent transfer to holding pressure.

As a result of solidification, the pressure in the molded part itself decreases over time, even when the holding pressure in the melt chamber is constant. Once it has dropped to the ambient value, the holding pressure phase is complete. As the curves in Fig. 6.3.9 show for pressures measured near the gate and far from the gate, this condition will be reached at different times for different locations.

The primary objective of the holding pressure phase is to compensate for the shrinkage in the molded part that occurs as a result of contraction during solidification or cross-linking by continuing to add melt. If the holding pressure is too low or the *duration of holding pressure* is too short, sink marks appear on the molded part, or the dimensions of the molded part are outside the specified tolerances.

During the entire phase in which pressure is applied by the injection unit, there is a force at the interface between the mold and machine nozzle trying to push the injection unit away from the mold. This force is overcome by the *nozzle contact force*. This is supplied hydraulically by the cylinders used to traverse the injection unit. This prevents separation of the nozzle from the sprue bushing during the injection and holding pressure phases.

Cooling Time

Even though the pressure in the mold may have fallen to the ambient value, the molded part usually cannot be removed because of insufficient dimensional stability. Part removal can take place only at a lower temperature. The additional time required for the part to remain in the mold until ejection takes place is called the *residual cooling time*. The total cooling time extends from the start of injection until the mold opens. On most types of machines, however, the residual cooling time is set. The cooling time required for a specific material and given molded part thickness can be calculated quite simply for the particular operating conditions. It will only be mentioned here that this time increases as the square of the part wall thickness.

Upon completion of the holding pressure phase, the screw recovers for the next cycle. This can take place with the injection unit in either the forward or the retracted position.

Screw recovery is not necessarily complete before the end of the cooling phase, as shown in Fig. 6.3.2, but can extend beyond the cooling time for thin-walled parts. An attempt is made, however, to keep the cooling time and the recovery time as short as possible in order that the cycle time not be extended unnecessarily.

Barrel Heating

This discussion of the conventional injection unit will conclude with a review of the various possibilities for heating the plastic. Two different approaches are commonly encountered. The first involves heating by means of *resistance-type heater bands* that enclose the barrel from the outside. The second is based on *liquid heating*. The barrel is divided over its length into separately controlled heating zones, three to five in small machines and more in larger machines. The temperature of each zone is sensed by means of a thermocouple in that zone. The temperature controllers installed in the machine control cabinet assure that the specified set points are maintained. The full capacity of the heating zones is usually needed only when the machine is heating up, since a temperature change can be brought about only by conduction during this phase.

The advantage of resistance-type heater bands (ceramic- or mica-insulated) lies in their relatively low price. In addition, they are extremely simple to install, and their heating capacity can be readily matched to requirements. It is likewise possible with such heating elements to generate quite high energy densities (usually, though, a specific heating capacity of approximately 4 to 5 W/cm^2 suffices). Finally, one further advantage that should be pointed out is that high temperatures can be achieved without problem; this means, among other things, without endangering personnel. The disadvantages lie in the relatively slow response of the system. As a result, too high a temperature is maintained for a relatively long time, especially when no heat is needed, since these heating elements cannot remove heat.

The major advantage of liquid temperature control is that with such systems it is also possible to remove heat from the melt. Thus, liquid temperature control is always to be recommended when temperature-critical molding compounds are to be processed. This is the case with all cross-linking systems.

6.3.2.2 Clamping Unit

After the cooling time has elapsed, the mold is opened for part removal/ejection. This requires that the two machine platens, between which the mold is located, be moved apart. This function is performed by the *clamping unit*, which is generally responsible for opening and closing the mold. In addition, the auxiliary mechanisms needed for part removal/ejection (cylinders for actuating ejector rods, compressed-air ejectors, etc.) are in most cases located in the clamping unit.

Only one machine platen is moved when opening or closing the mold. The platen closest to the injection unit is firmly attached to the machine frame and is thus immovable. It contains a circular opening in which the mold is positioned by its locating ring. The nozzle of the injection unit advances to the mold through the opening in the platen. This platen is called the *fixed*, or *stationary platen*, while the opposite platen is called the *movable platen.*

As a rule, the movable platen is guided on four tiebars during its motions. The speed of the opening and closing motions as well as the travel can be matched to the mold and production conditions.

After the two mold halves have been brought into contact, the *clamping force*, with which the halves are pressed against one another, is applied by the clamping mechanism. This force prevents the mold from opening as a result of the forces developed during the injection and holding pressure phases (*mold-separating force*).

Different designs, of which three types can be found on the market, are used to perform the functions discussed. These designs can be classified as:

- fully hydraulic,
- mechanical,
- hydromechanical.

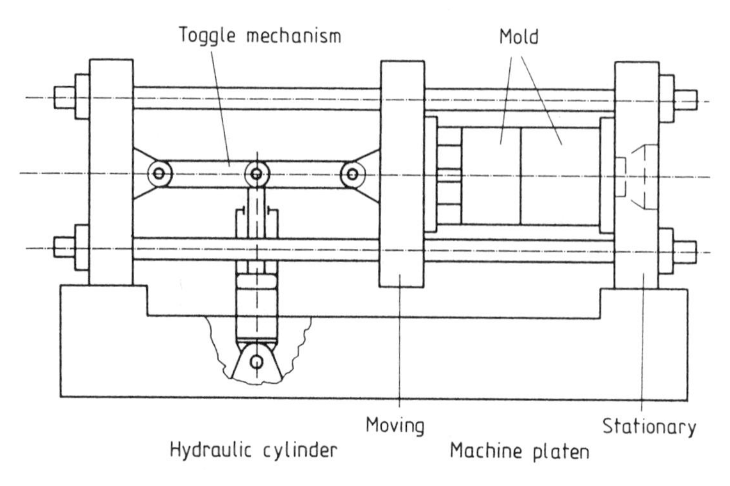

Figure 6.3.10 Toggle clamping unit.

systems, where, however, the so-called purely mechanical systems are actuated by a hydraulic cylinder. Here, only the two basically different systems will be discussed in brief:

- mechanically actuated toggle (Fig. 6.3.10),
- fully hydraulic clamping unit (Fig. 6.3.11).

Toggle Clamping Unit

In Fig. 6.3.10, a single-toggle system with fixed actuating cylinder is illustrated schematically as the simplest version of a toggle clamping unit. The clamping forces are generated by extending the toggle, which is normally actuated by an hydraulic cylinder. One of the major advantages of this arrangement is its force/speed characteristic. With the actuating cylinder operating at a constant speed, the moving mold halves achieve high speeds until shortly before closing simply because of the geometric relationship. The speed then decreases, while the attainable force increases sharply, so that high clamping forces can be generated with relatively small hydraulic cylinders. After the mold is closed, no further energy must be expended in order to maintain the clamping force during the injection and holding pressure phases. The maintenance requirements of these systems are very low, while the relatively high fabrication costs must be counted as one of the major disadvantages of toggle clamping units.

It must also be mentioned that a major advantage of toggle systems, the above-mentioned force/velocity characteristic, is no longer adequate in high-performance machines (packaging machines) to provide sensitive mold protection. In many cases, this characteristic must be improved by reducing the hydraulic pressure in the actuating cylinder.

Among other disadvantages of the toggle is surely that the clamping force changes as a function of the ambient temperature as a result of differential expansion of the tiebars. Monitoring or closed-loop control of the clamping force is quite involved, not only with regard to measuring the clamping force itself, but also with regard to adjustment of the mold mounting height.

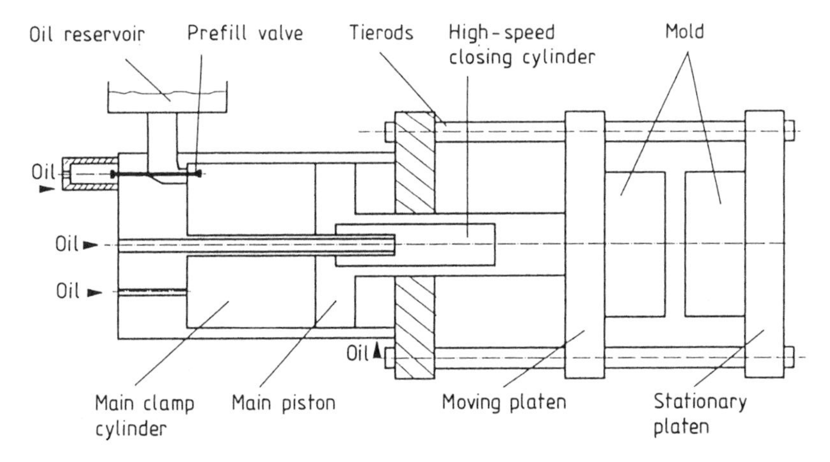

Figure 6.3.11 Fully hydraulic clamping unit.

Finally, compared to hydraulic clamping units, toggle systems occasionally have shorter opening strokes because of the design principle. Single toggles are employed in machines with clamping forces up to approximately 1000 kN (110 US tons) while larger machines with up to 4000 kN (440 US tons) clamping force employ double toggles.

Hydraulic Clamping Unit

The design of a *fully hydraulic clamping unit* is shown schematically in Fig. 6.3.11. The large cylinder required to apply the clamping force can be clearly recognized. In addition to this, there is usually a smaller jack cylinder for the opening and closing motions to eliminate the need for moving large quantities of oil under high pressure.

Because the stiffness (rigidity) of the oil column is less than that of steel, however, deformation during mold filling is greater with these clamping units than in mechanical systems. A further disadvantage of these systems is that fully hydraulic clamping units are slower than mechanical systems, since the oil column must be compressed when applying the clamping force.

There are also several disadvantages from the standpoint of energy consumption. Because of the larger volume of oil to be moved and the greater amount of energy required to compress the oil, these clamping units are at a disadvantage when compared with mechanical systems.

It can be assumed that, for opening and closing, the toggle system requires about 2/3 the energy of the fully hydraulic clamp. Since this has no effect on the remaining energy requirements, this means that the overall energy consumption of the fully hydraulic system is 10 to 15% higher.

Not to be overlooked is also the fact that hydraulic systems have only very few moving parts. This means reduced maintenance requirements. Finally, it must be mentioned that in these systems the clamping force can be adjusted very simply by means of appropriate valves in the hydraulics and maintained regardless of temperature fluctuations. Because of the relatively large surface area through which the force is introduced centrally in hydraulic clamping units, uniformly better stability of the movable mold platen results during operation.

6.3.2.3 Mold

The focal point of injection molding as a manufacturing process is the mold. It is not, however, part of the machine, but must be specially designed and constructed for each article to be produced. Errors in the design of the mold show up directly in the molded part quality or in the economics of the process.

In keeping with the variety of different injection molded parts, the number of possible design variations is unlimited, but all molds satisfy two basic requirements, which can be broken down further as shown in the following. The requirements involve

A) *Technological requirements*

- receiving and distributing the melt,
- forming the melt into the shape of the part,
- cooling the melt (thermoplastics) or introducing the activation energy (thermosets and elastomers),
- part removal/ejection.

B) *Design requirements*

- absorbing forces,
- transmitting motions,
- guiding the mold components.

The various functional groups must be matched to the part to be produced in accordance with these objectives. Using the injection mold shown in Fig. 6.3.12, the basic construction will be discussed.

Mold Construction

The majority of injection molds are constructed from *standard mold components*. These are design elements that are standard features of a mold and include bolts, guide bushings and pins, ejector pins and rods as well as return springs, sprue bushing, locating ring, and the plates that form the mold base. They are available from suppliers in assorted sizes designed to satisfy practical requirements. Mold plates are available in various grades of steel to permit hardening when necessary and must be specified during mold design on the basis of the requirements. Fig. 6.3.12 shows a mold with one *parting line* containing a flat part with attached *sprue*. The *movable-side mold plate* into which the *cavity* is machined can be easily replaced with another plate by loosening the bolts. In this way, various modifications of parts can be produced at little cost (e.g., switch plate with various hole arrangements). The molded part is then no longer gated via the central sprue, but via a runner system with pinpoint gates (Fig. 6.3.13). The stationary mold half (Fig. 6.3.12) consists essentially of the *base plate* (1), *mold plate* with temperature control (cooling) channels (2), *sprue bushing* (3), and *locating ring* (4) as well as the *guide pins* (5) and bolts with which the plates are located with respect to one another and held together.

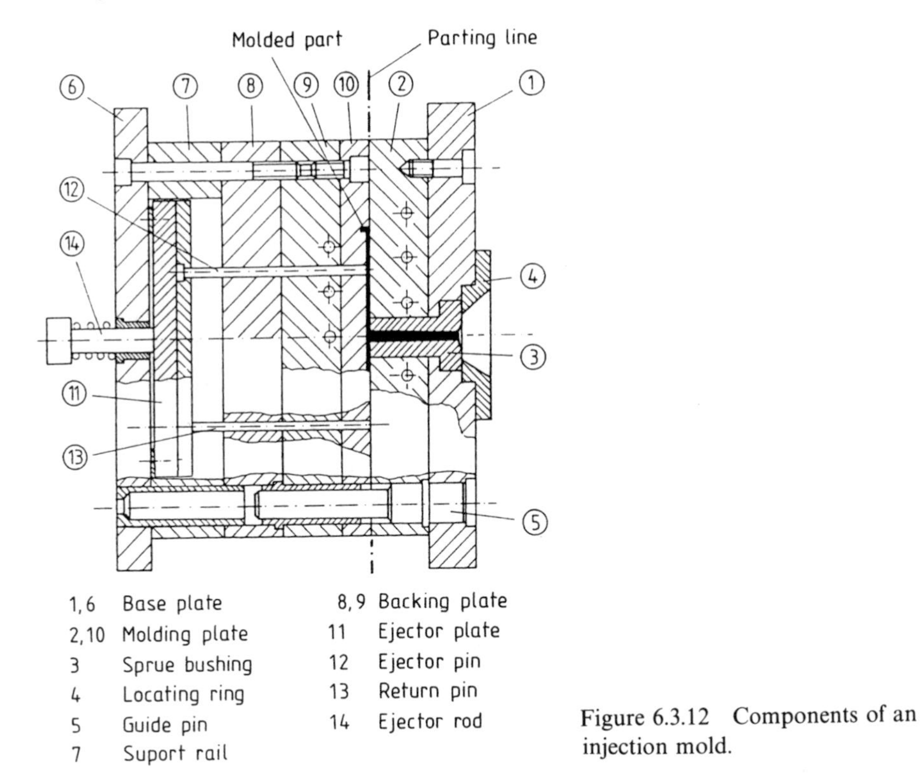

Figure 6.3.12 Components of an injection mold.

To locate the injection unit a spherical recess, the radius of which is matched to the radius of the nozzle on the injection unit, is machined into the sprue bushing. The location and number of temperature control (cooling) channels is likewise dependent on the molded part design and the material to be processed. If pressure transducers are to be employed in the mold, they should be installed in the stationary mold half if possible, since it does not move during production so that the necessary connecting cables from the sensor to the processing electronics are easier to route.

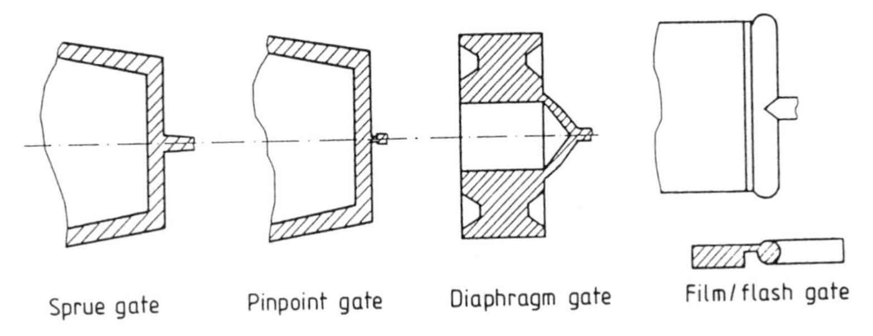

Figure 6.3.13 Gate types.

The two outer mold plates, the *base plates*, extend somewhat past the mold base to provide a clamping surface for the mold clamps with which the mold is fastened to the machine platens. In newer machines, automatic quick mold clamping systems are finding increasing use. With such systems, the mold is no longer held by means of bolts and mold clamps, but rather by means of hydraulic clamping mechanisms. These require, in turn, specially designed clamping plates.

The movable mold half consists of the *base plate* (6), *support rails* (7), a *bolster plate* (8) inserted to reduce deflection, and a *backup plate with cooling channels* (9) as well as the interchangeable *mold (cavity) plate* (10). The *ejector housing* with *ejector plates* (11), *ejector pins* (12), and *return pins* (13) as well as *ejector rod with return spring* (14) is shown as a moving design element. In principle, it is possible to combine into a single plate the backup/bolster plates and the mold (cavity) plate. The reinforcing bolster plate was incorporated to reduce deflection of the cavity, which can be significant if the support rails are spaced far apart.

After the molded part has achieved adequate dimensional stability (end of the cooling time), the mold is opened by means of the clamping unit. The ejector rod is now advanced toward the molded part with the aid of a hydraulic cylinder installed in the clamping unit. The ejector pins attached to the ejector plates press the molded part out of the cavity so that it falls from the mold. The ejection sequence is complete upon retraction of the ejector assembly. Retraction of the ejector is usually accomplished with the aid of a return spring. Positive retraction of the ejector rod is also employed occasionally.

With retraction by means of a return spring, jamming of the ejector assembly during production, e.g., because of dirt buildup, cannot be excluded. As the mold closes once again, the ejector pins protruding out of the cavity could then damage the often-polished mold surface or even break off.

To prevent this, return pins are often incorporated into the ejector assembly along with the ejector pins. They are located outside the cavity, however, and are flush with the parting line when the ejector assembly is retracted. Since they are longer than the actual ejector pins by the wall thickness of the molded part, they are the first to come in contact with the surface of the stationary mold half during closing in case the ejector has not been retracted. This returns the ejector assembly to the retracted position. In addition to the possibility for part ejection illustrated in Fig. 6.3.12, there are versions in which the *ejector pins* are replaced with *ejector blades*, *rings (stripper rings)*, or *plates (stripper plates)*. With rubber parts, part ejection often must be accomplished by hand or with the aid of compressed air because of the lack of rigidity in the molded part. The latter means also find application occasionally for thermoplastic or thermoset parts. In automatic production cells, part handling devices or robots are also employed to assist in part removal.

Runner System/Gating

When designing a mold, great importance also must be given to the rheological aspect of the design in addition to the purely mechanical aspect of sizing mold components and laying out the cooling system. This includes calculation of the flow conditions in the cavity, along with laying out the runner system and positioning the gates on the molded part. This latter point will be discussed briefly.

During injection the melt is forced through the *runner system* and into the molded part through the gate. The runner system can be heated or unheated.

With thermoplastics the melt remains fluid in the heated *hot runner system* and thus can be injected into the cavity during the next cycle. Advantages of this system include the low pressure requirement for part filling and the fact that the runner does not become scrap or have to be regranulated. Disadvantages are the higher costs of the hot runner system versus those for an unheated runner system, as well as the more complex design requirements and the need for peripheral instrumentation.

The counterpart of the hot runner when processing elastomers or thermosetting resins is the *cold runner system*. The lower temperature in this section of the mold prevents solidification of the molded compound as the result of cross-linking. In this way the material can also be injected into the cavity during the next cycle.

The molded part is connected to the runner system via the gate. It usually exhibits a significant reduction in cross section from that of the runner so that the molded part can be easily separated from the runner during part ejection. Some common types of gates are shown in Fig. 6.3.13. In addition to these, there are a number of systems of varying complexity depending on the number of cavities or molded part designs. For a more detailed discussion of this topic, the reader is referred to the literature.

Sensors

Because of the increasingly stringent quality requirements for injection molded parts, *pressure transducers* and *temperature sensors* are being installed more and more often in injection molds in order to permit monitoring of the cavity wall temperature as well as the course of pressure in the mold during filling. An illustration showing the course of the cavity pressure has already been presented in Fig. 6.3.9 for positions close to and far from the gate in a plaque filled from one end. The shapes of the curve are different for each part, however, and depend on the material and processing conditions. The shape of the curve can provide important information on potential sources of defects during production.

As was the case for the previously made statements with regard to machine and part size, the costs for a mold as well as its size and weight vary over a wide spectrum. Weights can range from 10^2 to 10^4 kg, while costs can vary from a few thousand DM for smaller molds to over 1 million DM for complicated molds used in the production of large high-quality parts.

6.3.2.4 *Mold Temperature Control*

Temperature control units are needed to remove heat from the molded part when processing thermoplastics and to provide temperature control for the injection unit when processing thermosets or elastomers. The number required for proper temperature control in molds depends on the complexity and size of the molded part, while for plasticating units the number of temperature control zones is the determining factor. Temperature controllers are heat-exchanging devices that can be purchased in various sizes having different heating capacities and pump ratings. As a rule, they are not part of the machine.

The normal heat transfer medium is either water (for temperatures less than 100°C) or oil (temperatures up to 300°C). The fluid moves in a closed circuit between the temperature control unit and the mold or barrel. The need to remove heat from this circuit is satisfied by injecting tap water, which flows through the temperature control unit as needed.

It is still possible to find occasional operations in which the thermoplastics processing molds are cooled only with tap water. This, however, has numerous disadvantages. Since the mold temperature is determined primarily by the temperature of the heat transfer medium and only minimally by the flow rate, there is hardly any possibility of establishing a mold temperature that is optimum for production.

Moreover, tap water contains traces of calcium carbonate, which can be deposited in the cooling channels of the mold, thereby impairing and possibly even preventing control of the mold temperature. In addition, the unnecessarily high water consumption cannot be justified from environmental or economic standpoints. Molds for processing thermosets or elastomers (rubber) employ electric heating for temperature control for the most part, and only seldom heat transfer fluids. With simple molded part geometries, indirect heating by means of *heating platens*, which are attached to the machine, is often employed. In complicated molds heat is provided by means of cartridge heaters. These permit heat to be introduced locally in differing amounts and are thus comparable to the cooling channels in a mold for processing thermoplastics.

6.3.2.5 Machine Frame and Control Unit

As a rule, the *machine frame* is of welded steel construction. It supports the injection unit and the clamping unit. In hydraulically actuated machines, the reservoir for the hydraulic oil and the pumps are also integrated into the machine frame. Occasionally, the control system and operator controls as well as mold temperature control unit are located in the machine frame.

Normally, the *control system* is located in a free-standing control cabinet next to the machine. In addition to the indicating instruments, it contains the electrical hardware, the temperature controllers, and the power supply system. While in older machines the set points had to be entered by means of thumbwheel switches or push buttons, the processing parameters are entered via a keyboard and video screen in state-of-the-art machines today. The microprocessor located in the control cabinet also controls the machine operating sequence, monitors process and production data, and stores data for documentation or use again at a future date. It is also usually possible via interfaces to control part- handling devices for removal of the molded parts.

6.3.2.6 Special Aspects of Rubber and Thermoset Processing

Since thermosetting plastics and rubbers differ from thermoplastics in that they solidify as a consequence of cross- linking upon the introduction of heat, certain steps of the injection molding process must be modified to accommodate this aspect. Special processing problems also arise from the viscosity characteristics of these materials, which differ from thermoplastics.

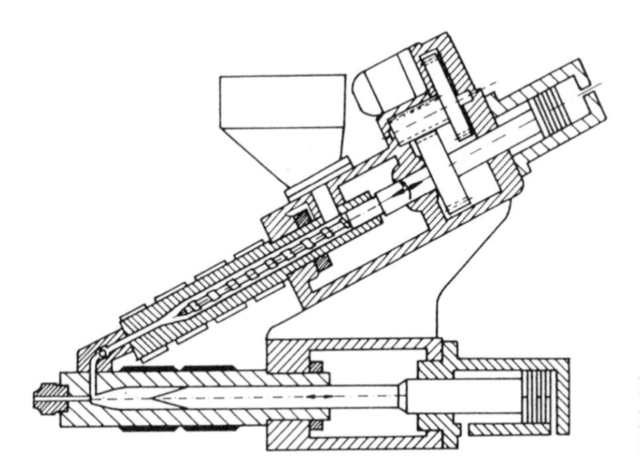

Figure 6.3.14 Screw preplasticating with plunger injection.

Basically, the process sequence corresponds to that for injection molding of thermoplastics. To avoid premature cross-linking in the injection unit, however, the temperatures there must be kept at about 80–100°C. Since it may be further necessary to remove frictional heat, the injection units are provided with liquid temperature control. Mold temperatures are usually 160–200°C, but must be individually matched to the material to be processed. There are noticeable differences between rubber and thermoset processing themselves as well as in comparison to thermoplastics processing with regard to the injection unit and screw design. Screws for processing cross-linking molding compounds are always shorter and have a more deeply cut feed zone than do screws for thermoplastics. In *thermoset processing*, the screws have only very little compression. Nonreturn valves are seldom used, since the dead corners here can lead to material hang-up. This material cures and prevents operation of the nonreturn valve. On the other hand, particles of cured material can break away and be injected into the molded part, where they can cause potential weak spots. Shutoff nozzles are never used in thermoset processing.

In *rubber processing*, screw injection alone is sometimes inadequate. Very often, rubber parts are thick-walled and very large. The injection units thus must be capable of providing large quantities of melt.

For this reason, a *screw and plunger arrangement* is often used (Fig. 6.3.14). Here, plastication and injection are accomplished by means of two separate cylinders. The rubber formulation, usually in strip form, is fed into the extruder section (the barrel of which usually has a feed pocket), plasticated, and fed into the injection chamber (pot) through a nonreturn valve. This assures very good *thermal and mechanical homogeneity*. The material is injected into the mold by a plunger through a shutoff nozzle.

In contrast to molds for thermoplastics, molds for rubber often utilize a three-plate design, i.e., they have two parting lines, with the runner system in one parting line and the molded parts in the other.

This design is chosen because rubber parts, due to a lack of rigidity, often cannot be ejected by the mechanical ejection mechanisms usually employed for thermoplastic parts. They must then be removed by hand. Since the molded part and runner system are located in two different parting lines, part removal is simplified as the parts are separated from the runners upon mold opening. For this reason, the mold and injection unit are often arranged vertically.

Additional problems arise when injection molding thermosetting molding compounds because of the very low viscosity of the melt compared to that of thermoplastics. Molding compound will flow into the smallest gaps in the parting line or between ejector pins and the mold frame. This leads to the formation of *flash*, which must be removed in a subsequent finishing operation.

An attempt is made to finish parting lines and openings in the mold with very smooth surfaces so that flash can be avoided. This, however, results in the problem that air trapped in the mold can no longer escape or can escape only with difficulty. It becomes compressed and leads to *burns* in the molded part or incomplete filling.

Through the use of special control strategies and design measures, however, the above-mentioned problems can be eliminated almost completely today.

6.3.3 Process Variations

Besides the basic process sequence described above for the production of injection molded parts, there are several process variations. Some of these can be performed on the same machines described above or on machines with only a few modifications.

Intrusion Molding

This process often finds application for the production of very thick-walled parts. Some of the melt is intruded into the mold directly during screw recovery using the pressure built up by the rotating screw. Complete filling of the cavity then takes place by means of the axial motion of the screw. The necessary holding pressure is also applied in this manner.

Injection Compression Molding

Injection compression molding can be performed on every injection molding machine with a hydraulic clamping unit. For toggle machines, a special sequencing program is necessary. In contrast to conventional injection molding, the melt is injected into a partially opened mold, which uses a shear edge design. The melt is formed into the molded part by the closing action of the mold. Application of holding pressure is not necessary. The minimal molded-in orientation represents a great advantage of this process over injection molding.

Multicolor Injection Molding

In multicolor injection molding, the molded part consists of two or more components of different materials or colors. A portion of the molded part is initially formed by the first component (Fig. 6.3.15). Following this, another mold cavity, which is filled with the next component, is connected by either rotating the mold or actuating a slide. Automobile tail lamp lenses are a typical application.

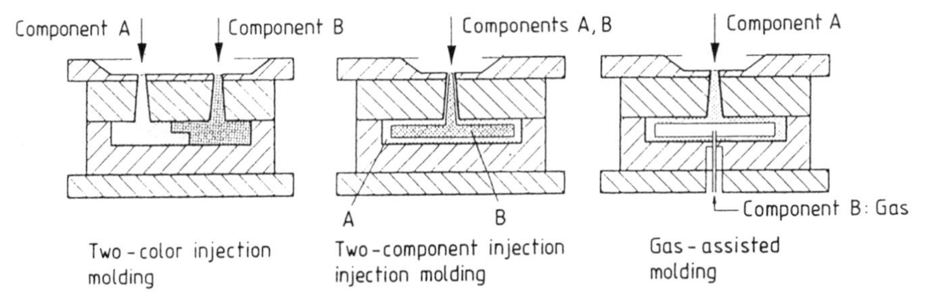

Figure 6.3.15 Process variations for injection molding.

Two-Component Injection (Coinjection) Molding

In contrast to the multicolor technique, two-component injection (coinjection) molding yields a sandwich structure in the molded part consisting of the different materials (Fig. 6.3.15). During the injection phase, the first component, which forms the smooth outer skin of the molded part, is initially injected into the mold. Following this, the sprue is connected to the second injection unit via a specially designed nozzle (valve) so that the second component can form the interior (core) of the molded part.

The material used for the second component can be either lower grade scrap or a material intended to impart specific functional properties (e.g., a conductive material used to achieve electrical shielding). The surface quality is determined by the skin material alone.

Gas-Assisted Injection Molding

Gas-assisted injection molding is comparable to two- component injection molding, with the core material being a gas (nitrogen) instead of a polymer (Fig. 6.3.15). With this process, thick-walled, highly rigid parts can be produced with a simultaneous savings in material and cycle time.

6.4 Production of Parts from Thermosetting Molding Compounds

6.4.1 Thermosetting Molding Compounds

Thermosetting molding compounds consist of resin and fillers; the resins—usually condensation resins based on phenol or melamine—have undergone preliminary condensation, i.e., they have already formed moderately long macromolecules. During the subsequent heating of primary processing, the reaction continues: The resin cross-links to form a nonmelting or insoluble thermosetting plastic. In addition to the above, molding compounds based on unsaturated polyester and epoxy resins are being employed increasingly today. Usually organic or inorganic fillers comprising 35 to 65 vol.% are added to the resins, along with up to 1% of lubricants and regulators. The most important filler is wood flour. It improves the toughness and stops crack propagation. Chopped fabric, glass fibers, stone flour, and

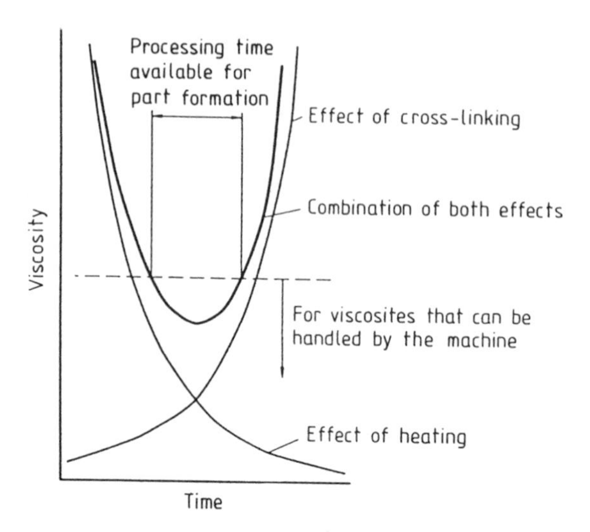

Figure 6.4.1 Viscosity curves of a compression molding compound.

similar fillers are also common. Condensation resins are prepared in the following manner: After the preliminary condensation, the resulting product is ground and mixed with the additives on a roll mill. After solidification, this product is ground once again. In contrast to the thermoplastics, thermosetting resins become solid not as a result of cooling, i.e., solidification, but rather as a result of curing. It can be seen in Fig. 6.4.1 that only a very limited period of time is available for processing, the processing window being determined by the reduction in viscosity as a consequence of heating on the one hand and the increase in viscosity as a consequence of cross-linking on the other. Once a certain degree of cross-linking has been achieved, the material can no longer be formed. Depending on the extent of preliminary condensation, a molding compound that cures more or less rapidly can be produced.

6.4.2 Compression Molding

The process sequence for the production of molded parts can be broken down as follows:

- preheating of the molding compound,
- metering and loading into the mold,
- melting (and, during compression, breathing to permit venting of the water vapor formed during the condensation reaction),
- molding (forming),
- curing,
- ejection,
- deflashing.

The individual process steps can be accomplished in the following manners.

Preheating

Preheating of the molding compound is accomplished:

– by means of *high- frequency*

An advantage here is development of heat from the inside out. A disadvantage is that this step can only take place outside the mold (for reasons of electrical insulation!). For better heat conduction, the molding compound should be tableted.

– by means of *conduction of heat* from the hot walls of the mold

An advantage here is that the transfer of the molding compound from a separate preheating station into the mold is eliminated. A disadvantage is the additional idle time needed to heat the molding compound by conduction.

– in an *oven*

This is possible only up to a temperature where sticking does not yet occur, for which reason this method is employed only in automatic machines (compare Fig. 6.4.2) because here the other methods are not applicable.

– by means of *friction*

This method is employed in transfer molding (compare Fig. 6.4.3) and in injection molding.

Metering and Loading

Metering in automatic compression molding machines (compare Fig. 6.4.2) usually takes place by volume. With manually loaded molds and for *transfer molding*, metering is also possible by weight or number of tablets (preforms) with properly prepared material. Loading in automatic compression molding machines is accomplished by means of gravity.

Melting

Melting of the molding compound is possible

– in the mold (primarily in automatic compression molding machines),

– in the screw or in the transfer chamber of the transfer molding press (or in a shear gap).

Molds

Depending on the manner of introducing the heat needed for melting, the part-forming process occurs simultaneously or subsequently by means of compression as the mold closes or by means of injection from the above-mentioned plasticating devices (transfer chamber, barrel). The molds are essentially the same as injection molds, except that they are always heated. Often, heating is accomplished indirectly by means of conduction from the heated press platen.

Curing

Curing is the result of a chemical reaction. The heat required can be introduced through the hot cavity walls or can still be present from the melting process in the plasticating

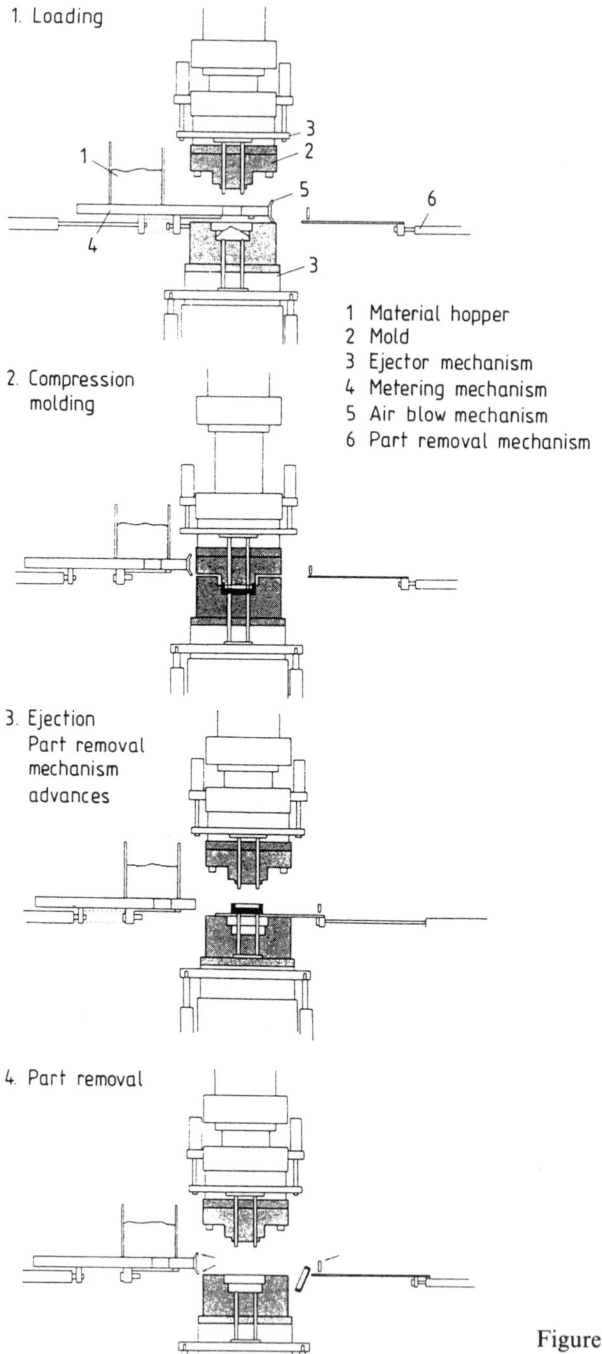

Figure 6.4.2 Automatic compression molding machine for producing parts in compression molding compounds.

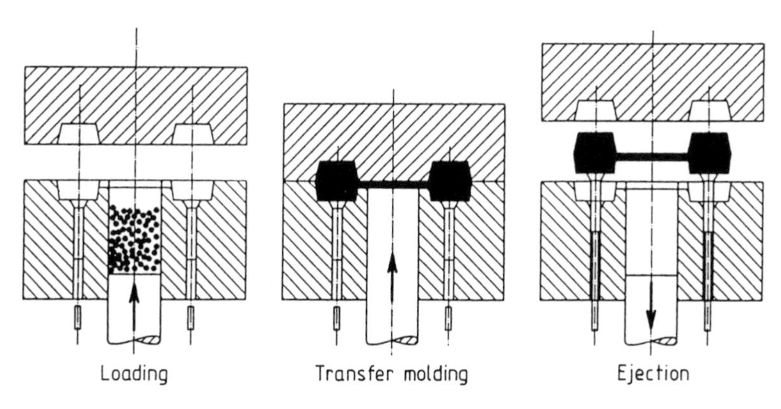

Figure 6.4.3 Transfer molding.

mechanism. Just as the cooling time is the cycle- and thus cost-determining factor in the processing of thermoplastics in injection molding machines, the curing time plays a dominant role in the compression molding process.

Ejection

As in injection molding, ejection of the molded parts can be accomplished by means of an ejection or stripping mechanism, or the parts can be removed by hand. The subsequent cleaning of the mold cavity with the aid of an air blast should not be considered unimportant.

Deflashing

While thermoplastic melts solidify especially rapidly in gaps in a cooled mold, compression molding compounds become especially fluid in gaps in heated molds and penetrate deeply. The result is flash, the removal of which requires a subsequent operation that can bring into question the economics of this method in spite of the inexpensive materials.

Process Variations

Based on the above possibilities, three methods are found in actual practice.

Compression Molding

Fig. 6.4.4 shows the two common types of molds. The positive mold has a tightly sealing parting line. Very exact metering of the molding compound is required since excess material cannot escape and insufficient material results in voids and pores. Positive molds are thus favored for large parts in single-cavity molds, since in such cases gravimetric metering is simpler.

The flash mold requires less accurate metering. Excess material is always loaded, which can then escape via the flash relief. Accordingly, this type of mold is suited for small parts and automatic compression molding machines in particular.

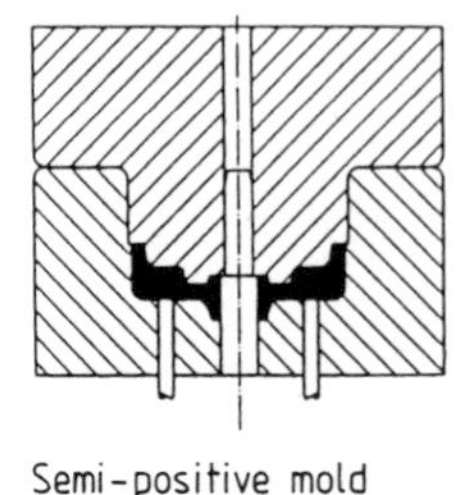

Semi-positive mold
(Landed positive mold)

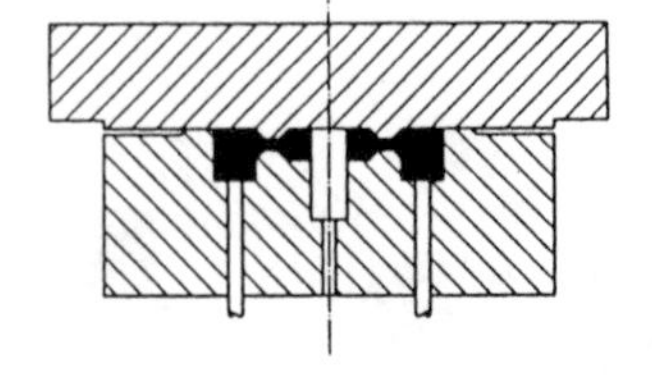

Flash overflow mold

Figure 6.4.4 Compression molds for processing thermosetting molding compounds.

Transfer Molding

While in compression molding the material in the mold is plasticated by heat from the cavity wall, plastication of the molding compound during transfer molding takes place in a separate chamber (Fig. 6.4.3). As in injection molding, the plasticated compound is injected into the mold cavity via a sprue. Another possibility, however, is heating the tablets (preforms) of molding compound prior to placement in the transfer chamber, e.g., by means of high frequency. The process sequence is as follows:

- possibly high-frequency preheating of the molding compound,
- placement of the molding compound into the hot transfer chamber and plastication,
- injection of the plasticated molding compound into the hot mold cavity,
- curing under the pressure of the transfer ram,
- ejection/removal of the hot transfer molded part,
- deflashing and mold cleaning.

The advantages of transfer molding over compression molding include better and more rapid heating; better venting (via the parting line), which makes a separate venting step unnecessary; more exact metering of the molding compound; and easier deflashing because of the thinner flash. The improved homogeneity resulting from passage of the molding compound through narrow flow channels in particular leads to shorter curing times and good quality.

The anisotropic mechanical properties resulting from the orientation of filler particles and the somewhat higher material consumption because of the runners are disadvantages. The process is especially well-suited for production of small and medium-sized parts. Transfer molding competes with automatic compression molding and especially injection molding, by which it is being displaced more and more.

Injection Molding

With regard to the machines used and their controls, the injection molding of thermosetting molding compounds is comparable to the injection molding of thermoplastic molding compounds. The plasticating screw, however, must be modified. The screw, barrel, and nozzle must not exceed a temperature of about 150°C in order to prevent premature curing of the molding compound in the barrel. The necessary close temperature control is best achieved by means of liquid heating. The screws are simple feed screws to avoid excess heating due to friction. Because of the associated dead corners where molding compound could cure, nonreturn valves and shutoff nozzles are not suitable. The mold itself has a temperature over 160°C. The advantages and disadvantages correspond largely to those of transfer molding. Nevertheless, injection molding is increasingly displacing the other methods, because with it automatic operation is more readily achieved, which makes the process more economical.

6.5 Foaming of Plastics

Foamed materials, i.e., materials with a cellular structure, can in principle be produced from almost every resin. The application of such methods, however, often involves considerable process difficulties and/or costs, so that to date only a small number of the possible methods have been successful in the foam market. What reasons favor the use of foams instead of the solid polymer matrix:

pro: low thermal conductivity
low weight in conjunction with high mechanical strength
material savings
sound attenuation
mechanical damping
good machineability

con: special processing technique
effects involving polymer formation/foaming reaction

The field of application for such materials results directly from this general property profile (Table 6.1).

Table 6.1 Fields of Application for Foamed Plastics

Application	Example
a) Technical molded part	Housings Enclosures Molded parts
b) Semi-finished goods	Mattresses Slabs Profiles Film/sheet
c) Insulation	Chiller panels Foam backings Sound-absorbing panels Heat shields
d) Packaging	Shipping packaging Fill

Process	Activation	Reaction type	Blowing agent	Mold	Examples
a) Injection molding	Thermal	Softening/cooling	Chemical	Closed	PVC, PE
b) Extrusion	Thermal	Softening/cooling	Chemical	Open	PVC, PE
c) Multi-component system	Mixing	Polyaddition	Chemical Physical Mechanical	Open/ closed	PUR
d) Multi-component system	Thermal	Polymerization	Chemical	Closed	PA, EP
e) 2-step sintering process (foam beads)	Thermal		Physical	Closed	EPS (Styrofoam)

Figure 6.5.1 Foamed polymers.

Production of a foamed material with an *open* or *closed cell structure* can be achieved economically via one of the methods shown in Fig. 6.5.1. Methods "a" and "b" involve the processing of *thermoplastics*. In addition to the polymer, an expandable gas is generated with these methods, which makes it possible to obtain a thermoplastic foam in the mold or in the calibration section. As with normal thermoplastics processing, it is also possible here to achieve increased thermal or mechanical properties through cross-linking.

The method for producing foamed beads (e) is a special sintering method [polystyrene foam beads (Styrofoam)]. This technique is employed in the production of molded parts for insulation and packaging purposes. The method involves a *two-step sintering process* in which expandable polystyrene beads are processed into prefoamed beads. In the second step, the beads are allowed to foam fully while confined in a mold and weld together to form the molded part.

The most interesting process for the production of foamed plastics from the technical and economic standpoints involves the processing of reactive *multicomponent resins* (processes "c" and "d"). The most important representative of this group of plastics is polyurethane (PUR). The spectrum of material properties is shown in Fig. 6.5.2 in relation to the properties of the solid material and the attainable density of the foamed part.

Since polyurethane foam is at present the most important foamed plastic, the mechanism of foaming and the possibilities for processing such plastics will be discussed in the following using multicomponent foams as an example.

6.5.1 Foaming of Reactive Resins

A prerequisite for the foaming of plastics is that the polymer be in a flowable condition during the foaming. When foaming thermoplastics, this is achieved by heating the polymer

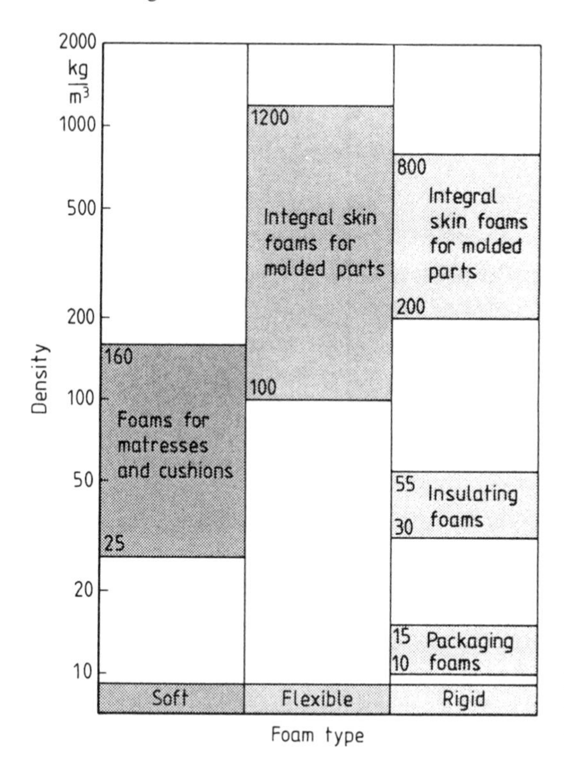

Figure 6.5.2 Fields of application for polyurethane foams with various densities.

matrix to temperatures above the *softening point*. The flowable condition is necessary first of all to permit formation of *bubbles* and the *cell membranes*. In addition, material transport through the growing volume takes place at the surface and in the interior of the foaming plastic.

When processing low-viscosity reactive resins, the prerequisite for a flowable condition is satisfied in that the base components as well as the mixture formed are highly fluid substances. The transition from the fluid mass to a solid resin can in principle be achieved in the following two manners:

a) The mixture reacts to form a high-molecular-weight and/or cross-linked polymer. The bubbles formed are fixed in place by the chemical reaction at elevated temperatures.

b) The mixture reacts to form a high-molecular-weight thermoplastic with the requirement that the reaction temperature be below the softening point of the polymer formed.

When producing polyamide (nylon) foam, this temperature must be significantly below the crystallite melting temperature of the polymer so that the bubbles are fixed in place by solidification of the matrix structure.

A prerequisite for the formation of bubbles is that an additional component be present in the reaction mixture in gas form. This component is generally called a blowing agent. In addition to the blowing agent itself, further additives such as foam stabilizers and surface-active substances as well as nucleating agents are necessary in order to achieve a uniform foam structure. The following is a list of various foaming mechanisms employed at present, along with the reasons for foaming:

a) *Chemical foaming method*

- reaction of the blowing agent with one reagent, e.g., foaming of polyurethane with water;
- thermal decomposition of the blowing agent.

b) *Physical foaming method*

- vaporization of a low-boiling point liquid by means of an exothermic reaction, e.g., CFC (Freon) as a blowing agent for foaming polyurethane;
- vaporization of a low-boiling-point fluid by heating a thermoplastic, e.g., prefoaming of "inoculated" polystyrene pellets to produce EPS.

c) *Mechanical foaming method*

- dispersion of gas and expansion upon reduction of pressure, e.g., charging highly reactive resins with air;
- high- pressure injection of a compressed gas and expansion in the mold, e.g., the frothing method for polyurethane.

The manners of *polymer formation*, selection of the foaming method, and possible variation of the mixture ratio with multicomponent resins result in such a wide spectrum that polyurethane can be called a *custom-made material.* A prerequisite for the production of satisfactory foams of polyurethane or other systems is that, first, the individual components be mixed with one another as well as possible. After the starting components are mixed, an

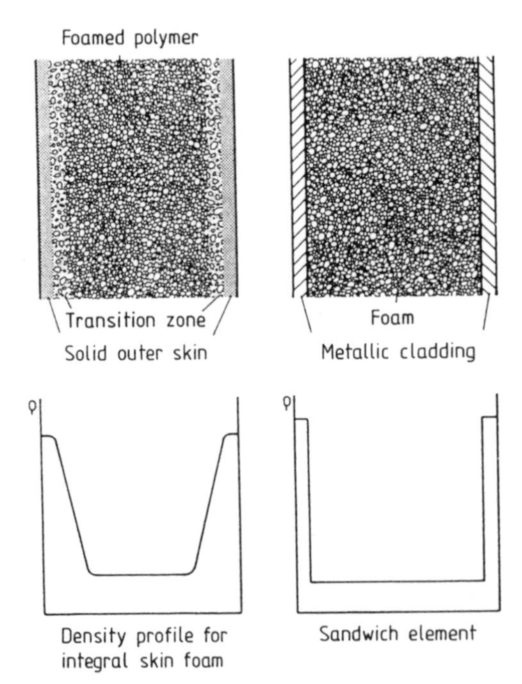

Figure 6.5.3 Foam structures as engineering materials.

exothermic reaction takes place with simultaneous formation of the foam structure. Three phases of foam formation [bubble creation (nucleation), bubble growth, bubble retention (fixing)] are essential for development of the foam structure, which is characterized by the size, number, and geometry of the bubbles. By selecting a particular blowing agent and specifying the physical boundary conditions, the process can be controlled in such a manner that polyurethane parts can be produced with a cellular core, a largely noncellular skin region, and, in ideal situations, a parallel density distribution across the part thickness. These are called *integral skin foams* or foams with an integral density distribution. Such foams are employed today for mechanically loaded parts because of the stiffness that can be achieved at low weights as a result of the sandwich structure (Fig. 6.5.3).

6.5.2 Processing of Low-Viscosity Reactive Resins

Whether a plastic is foamed, solid, or produced with an integral density distribution across the part thickness is not only a question of the materials, but it also can be influenced to a large extent by the accompanying process technology. This will be discussed in the following using the production of molded parts by means of *reaction injection molding* (*RIM*) as an example.

The RIM process is characterized by the following combination of process steps:

- metering of the reagents;
- mixing;
- injection into the mold and filling of the cavity;
- reaction in the mold, with formation of the foamed or solid part;
- ejection/removal of the molded part.

The first four steps are directly coupled to one another by the process because of the highly reactive nature of the materials to be mixed (reaction times down to 1 s).

The course of the reaction can be described on the basis of the following processes occurring in the mold (Fig. 6.5.4). Initially, the mixture of reagents is very low in viscosity and then becomes increasingly viscous as a result of the reaction until the *gel point* is reached. At this point, there is a temperature rise, which directly affects the foaming, as a result of the polyaddition reaction that yields polyurethane.

The processing of reactive *multicomponent resins* requires special processing technology. The core of this technology is the production of the mixture from at least two reagents.

A basic prerequisite for the production of a good mixture is delivery of the reagents (identified here as A and B) at a constant rate (Fig. 6.5.5). This can be accomplished with sufficient accuracy by various machine types at freely selectable values of pressure and temperature. The material is mixed either by converting the stored pressure energy into velocity in a type of *impingement mixer* or by pumping the material into a *mixing chamber*, where it is homogenized by agitators. Following this, the material is allowed to settle to such an extent that the reaction mixture enters a mold or the surroundings as quietly as possible, i.e., without entraining air.

The type of mixer also directly determines the design of the metering devices. If the material is mixed by an agitator, the machine must operate at only one pressure level sufficient to

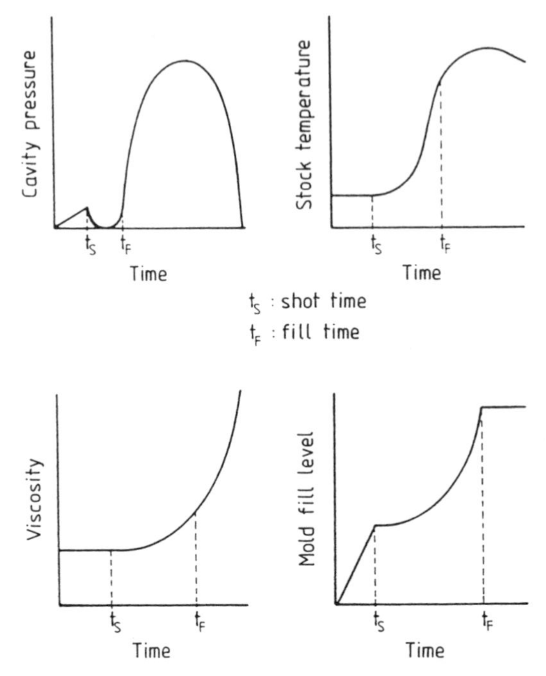

Figure 6.5.4 Curves showing changes in physical properties during production of PUR foam parts.

convey the material through the lines. For this reason, agitator-type machines are also called *low-pressure machines* (Fig. 6.5.6).

To achieve a homogeneous mixture in an impingement mixer, the system must operate at a higher pressure level. Systems of this kind are thus called *high-pressure systems.* In addition, selection of the mixing method predetermines the possibilities for further processing. *Low-pressure mixing heads* (*agitator mixing heads*) must be cleaned or flushed after delivery of the mixture. The consequence is that such mixers cannot be attached directly to a mold. Accordingly, only semi-finished goods are produced or open molding methods (injection into an open mold, coating) are employed. When high-pressure systems are used, the mixing heads are cleaned mechanically by pistons upon completion of the shot, resulting in a process comparable to injection molding with the mixing head attached directly to the mold. Depending on the reactivity and requirements for the molded part, the material is injected into either an open or a closed mold (Fig. 6.5.7), where it foams. Rapidly reacting systems, integral skin foams, or high-density foams are processed almost exclusively in closed molds. Tooling design is similar to that for injection molding, but when producing foam plastics via the RIM process only one gate can be used. Nevertheless, the low viscosity of the material permits production of thin-walled parts with large areas, i.e., long flow paths that are not possible when injection molding thermoplastics via a single gate. Molding with open molds is often more advantageous with RIM, since it is not necessary to design and machine an inlet channel. Use of this method is limited at present, however, to slower systems and to the production of foam-backed parts (e.g., instrument panels for automobiles) or products with additional reinforcements.

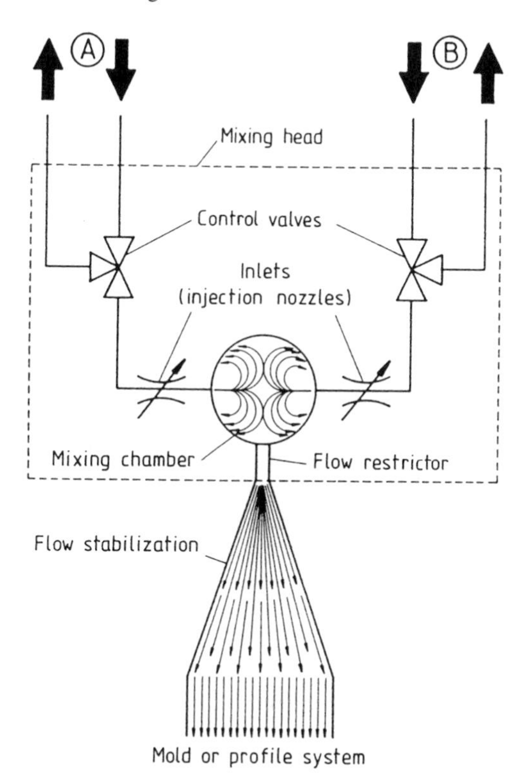

Figure 6.5.5 Mixing model for multicomponent processing.

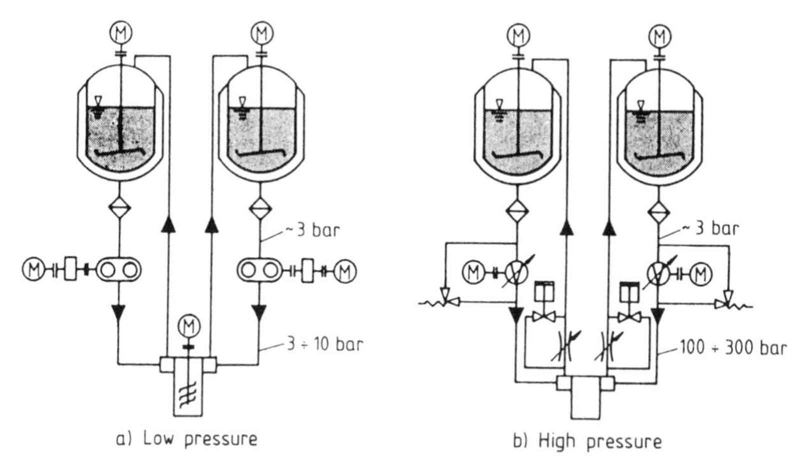

Figure 6.5.6 Process schematics for multicomponent metering and mixing systems.

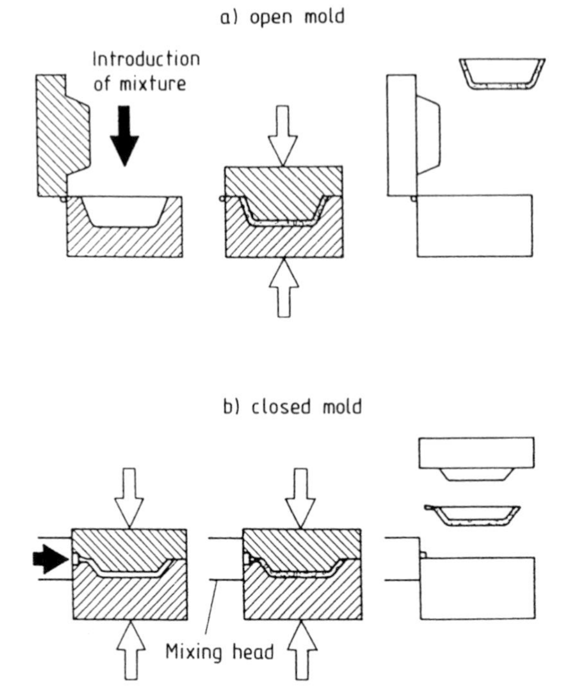

Figure 6.5.7 Production of foamed reaction molded parts.

A combination of one *metering device* with several molds is often employed with this method, which is seen relatively seldom in injection molding. In addition, foamed semi-finished products can be produced continuously with reactive systems (Fig. 6.5.8). Here, the reaction mixture formed in the mixer does not need extensive calibration but, because of the low viscosity, only a type of *guide*, which in simple cases can be provided by shaped paper. The block foam produced in this manner, for instance, is used in large amounts in the upholstery

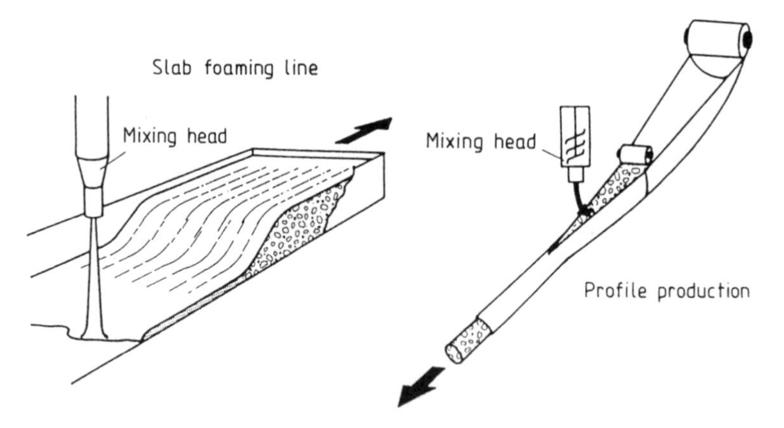

Figure 6.5.8 Continuous production of foamed semi-finished products from reaction resins.

industry as well as for insulation. Profiles have a wide range of application, extending from sealing profiles to high-grade technical products such as window profiles.

6.6 Reinforcement of Plastics

It has been known for some time that the properties of materials can be improved dramatically by incorporating specific reinforcements. Clay tiles reinforced with straw are an old example, as is steel-reinforced concrete. In both cases, the fibers (straw, steel) transmit the forces, while the surrounding material (clay, concrete) fulfills other functions, such as providing support, acting as a barrier, etc. This combination yields an *anisotropic material*, i.e., the material properties are not the same in all directions. Usually, they are better in the direction of the fibers than perpendicular to them. It thus also seemed natural to improve the properties of plastics by means of reinforcing fibers. This led to the development of long-fiber-reinforced plastics.

The term *long-fiber-reinforced plastics* is used to describe composites consisting of reinforcing fibers embedded in a plastics matrix. In recent years, these materials have gained continually in importance, since they offer an especially light-weight alternative to metallic materials. The specific stiffness and strength of various *fiber-reinforced composites* are shown in Fig. 6.6.1, along with those of metals for comparison. A premium for this favorable strength/weight relationship will only be paid, however, if a cost savings can be expected somewhere else. The rudder unit of the Airbus 300, which represents the largest structural element of

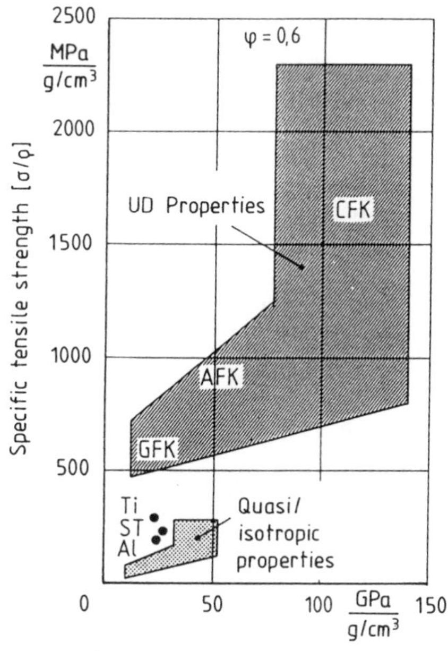

Figure 6.6.1 Specific properties of fiber-reinforced plastics laminates (composites).

fiber-reinforced plastics, is the best-known application. Traditionally, composites are employed for constructing pipelines and large containers or vessels. These applications exploit the good corrosion resistance of the material.

Another field of application can be found in the area of antennas and measuring instruments, where the negative coefficients of thermal expansion of various types of fibers, for instance, carbon fibers, are employed in the construction of extremely stiff, thermally stable structures. The extensive design possibilities are also employed in ship building and model construction. In general, fiber-reinforced composites are used for large flat parts and rapidly accelerated components.

6.6.1 Materials

As has already been mentioned, long-fiber-reinforced plastics consist of a *reinforcing fiber* and a *matrix*. The fibers absorb the tensile and compressive forces applied to the components. Understandably, they can only transmit forces in the direction of the fiber so that the *orientation* in the part is of decisive importance. The function of the matrix is to hold the fiber structure together and in this way transmit the shear forces between the individual fibers.

Glass fibers, carbon fibers, and even fibers of polymeric materials, such as aramid and polyethylene fibers, are employed. Glass fibers are the least expensive and most commonly utilized reinforcing fibers. They offer good *strength* and *elongation*. Carbon fibers exhibit both high strength and modulus, but at a rather high price. Aramid and PE fibers have almost the same strength as carbon fibers but a lower *modulus of elasticity*. They possess a high energy-absorbing capacity under impact, but relatively poor adhesion to resin. The properties of the various fibers are summarized in Fig. 6.6.2. It is also possible to combine different fibers with one another.

Property	Units	Glass fiber			Carbon fiber				Aramid fiber		PE fiber
		E	R/S	C	HT	HST	IM	HM	Normal	HM	
Tensile strength	GPa	2.3	1.9–3.0	2.1	2.7–3.5	3.9–7.0	3.4–5.9	2.0–3.2	2.8–3.0	2.8–3.4	2.6–3.3
Modulus of elasticity	GPa	72–73	86–87	71	228–238	230–270	280–400	350–490	58–80	120–186	87–172
Elongation at rupture	%	2.2–3.2	2.8–3.6	2.3	1.2–1.4	1.7–2.4	1.1–1.9	0.4–0.8	3.3–4.4	1.9–2.4	2.7–3.5
Spec. tensile strength	GPa·cm³/g	0.9	1.13–1.23	0.9	1.5–2.0	2.2–3.0	2.0–3.1	1.1–1.7	1.9–2.2	1.9–2.3	2.7–3.4
Spec. modulus of elasticity	GPa·cm³/g	27.7–28.2	34–34.9	29	127–134	127–150	160–200	190–260	40–56	83–127	90–177
Filament diameter	μm	3–25				5–7	5–7	6.5–8.0	12	12	27–38
Density	g/cm³	2.6	2.5–2.53	2.45	1.75–1.83	1.78–1.83	1.73–1.8	1.79–1.91	1.39–1.44	1.45–1.47	0.97
Coeff. of thermal expansion	10^{-6}/K	5	4	7.2	−0.1–−0.7		—	−0.5–−1.3	−2.0–−6.0		< −9

Figure 6.6.2 Typical properties of various fibers.

Material	Use temperature °C	Elongation at rupture %	Tensile strength MPa
Polyester (UP)	<80	1...3	30...60
Vinylester (VE)	<100	4...6	80
Epoxyresins (EP)	120...230	2...>10	50...100
Phenolicresins (PF)	150...230	0.3...1	20...60
Polyimides (PI)	200...310	2...>10	80...100
Polypropylene (PP)	<100	20...800	30...40
Polyamide/nylon (PA)	70...130	40...500	40...60
Polyethersulfone (ES)	200	40...80	80...90
Polyetheretherketone (PEEK)	240...260	40...80	80...90
Polyphenylenesulfide (PPS)	240	—	75

Figure 6.6.3 Material properties of common matrix materials.

For instance, a hybrid of carbon fibers and aramid fibers is extremely strong (carbon fibers), yet retains good residual elongation (aramid fibers, high degree of elongation) when the carbon fibers break. All fibers are given a *coating* during production to protect them against damage during subsequent processing. This coating is then usually replaced by a coupling agent, which is intended to provide good wetability of the fibers and thus good contact with the matrix. In addition to continuous fibers (*rovings*), flat products such as *woven* and *nonwoven fabrics* are employed. Nonwoven fabrics are available in a variety of forms. They can be based on statistically distributed continuous or short fibers as well as on *uni*- or *bidirectionally* oriented continuous fibers.

The matrix usually consists of a *thermosetting resin*, such as polyester, epoxy, or phenolic resin. They differ in terms of strength, maximum permissible elongation, service temperature, shrinkage upon processing, and chemical resistance (see also Fig. 6.6.3).

Often, fillers and processing aids, such as *thickening agents*, *accelerators*, *colorants*, and *stabilizers* are added. All thermosetting matrixes have a very low viscosity during processing, which promotes coating of the fibers.

Often, a preloaded product consisting of fibers and matrix, the so-called *prepreg*, is processed. To facilitate handling of the prepregs, *hot-curing resins* are employed that are extremely viscous at room temperature and do not drip. In this way, the processor receives from the supplier a predetermined, constant ratio of fiber to resin volume and good impregnation of the fibers. Prepregs can also be produced from unidirectional nonwoven fabrics of continuous fibers as well as woven fabrics. Products with large surface areas are produced from prepregs extremely often.

At the same time, fiber-reinforced composites based on thermoplastic resins continue to gain in importance. Among their advantages over materials based on thermosetting resins is the higher elongation at rupture (see Fig. 6.6.3) (polyether sulfone, polyether ether ketone), which results in better impact strength. In addition, they exhibit poor flammability. The ability to melt the matrix again permits parts to be welded to one another. Better recyclability is also listed as a further advantage. The more involved molding processes and poor wetting of the fibers create problems, however.

6.6.2 Processing Methods

The processing methods for fiber-reinforced composites have a decisive effect on the fiber content and fiber orientation in the resulting material. A variety of processing methods is available for different part geometries, fiber types, and matrix resins.

6.6.2.1 Hand Lay-Up Method

The *hand lay-up method* (Fig. 6.6.4) is the simplest processing method. It always requires a mold of wood or other material that determines the geometry of the part to be produced. The laminate is applied to this mold. It consists of the matrix (usually polyester resin) and the reinforcing fibers. The resin is applied to the mold first. Next, mats of chopped strands, woven or nonwoven fabric are worked in with a brush or laminating roller in order to achieve good impregnation. In this manner, several layers of fibers are built up on one another, i.e., laminated, until the desired wall thickness is reached. A good surface is obtained by applying a gel coat to the mold. This is a layer of pure resin (usually filled) that is intended to cover and conceal the fiber texture. Hand lay-up is a very labor-intensive method but does not require much capital investment, which makes it suitable for the production of very large and complex parts in small quantities. It is not possible, however, to exactly maintain a specified fiber orientation. In addition, the fiber content is usually less than 45 vol.% so that high-performance composite parts cannot be produced via hand lay-up.

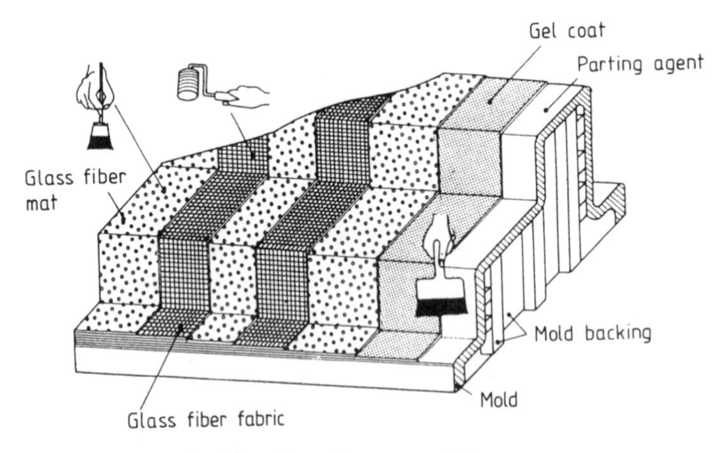

Figure 6.6.4 Principle of hand lay-up molding.

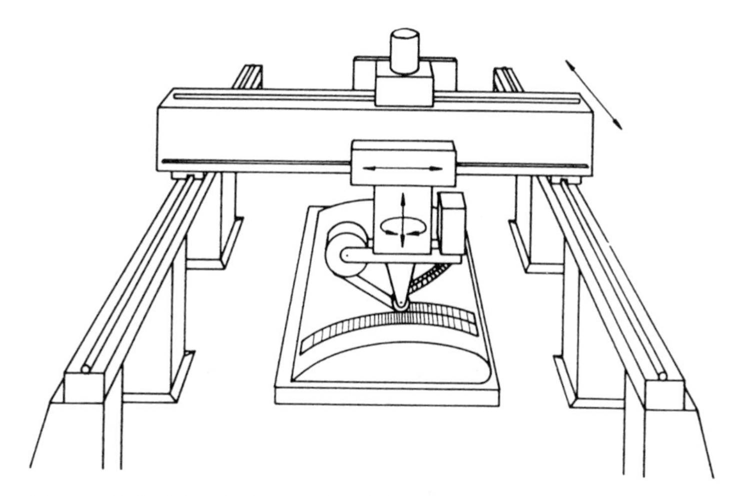

Figure 6.6.5 Tape laying with a bridge-type robot.

Typical applications include the production of large parts, such as boats, or large containers and prototypes as well as bonding and repair of fiber-reinforced composite components. The hand lay-up method is employed often in mold making, since even very complex geometries can be achieved.

6.6.2.2 Prepreg Processing, Tape Laying, Autoclaving

If parts are to be produced in compliance with very high quality standards, *prepreg processing* is employed. Prepregs are fibers preimpregnated with matrix material. These are cut to size and placed in the mold manually or mechanically in accordance with the desired fiber orientation.

Placement can also be automated (Fig. 6.6.5) with the aid of a tape layer (bridge-type robot). Following this, the laminate on the mold is covered with a sheet and cured under pressure in an autoclave at elevated temperatures. By applying a vacuum, the air between the individual prepreg layers is removed. The positive pressure in the autoclave then compresses the laminate uniformly. The highest quality parts are achieved in this manner, since the fibers can be placed exactly in the direction of loading.

6.6.2.3 Spray-Up Molding

Spray-up molding is suitable for small to moderate quantities of molded parts (Fig. 6.6.6). The piece of equipment used is a spray gun (usually manually controlled), the purpose of which is to supply, mix, and spray onto a mold the resin, curing agent, accelerator, and chopped fibers. The continuous fibers that are fed in are chopped to the desired length by the cutter and sprayed out with the aid of compressed air. To compress and remove air from the laminate, a laminating roller can be used, as in hand lay-up molding, or a plastic film can be placed over the laminate on the

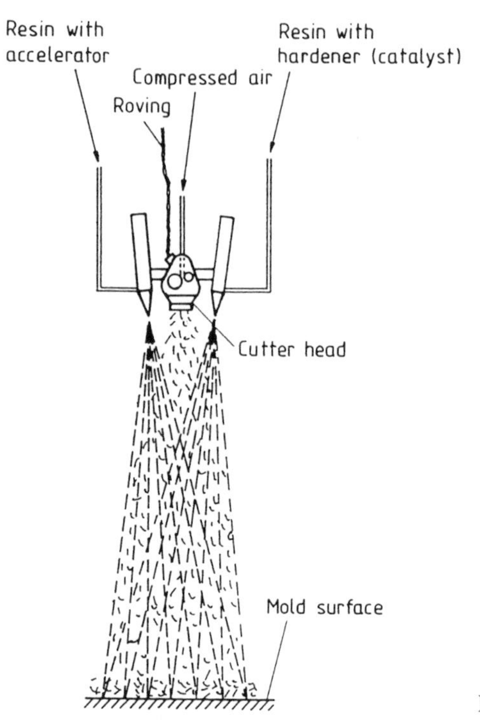

Figure 6.6.6 Fiber spray gun.

mold and evacuated so that the air bubbles escape as the ambient pressure compresses the laminate. In order to achieve a good surface, a gel coat is applied to the negative mold. This process employs exclusively polyester resin, which, however, releases styrene into the surrounding air as the resin is atomized. Typical parts produced via the spray-up method include bath tubs, containers, roofing elements, and swimming pools.

6.6.2.4 Filament Winding

Filament winding is a largely automated method for producing hollow parts of fiber-reinforced plastics. In this process, the reinforcing fibers are wound around a mandrel (core). Depending on the geometry of the mandrel, it is either used again, remains in the part as a lost core, or is produced from a soluble material. Lost cores that fulfill a function in the part are called *liners* and protect against aggressive substances or act as a *diffusion barrier* in gas containers. The most commonly employed winding method operates on the *lathe principle*, as shown in Figs. 6.6.7 and 6.6.8. The rovings are drawn from a spool, with the filament tension held at a constant value regardless of the filament velocity. The *filament tension* is the deciding factor for the fiber content of the part. In the *impregnation bath* the rovings are opened up, wetted with resin, and then thoroughly impregnated in a tumbling unit. With sensitive fibers, e.g., high-modulus carbon fibers, the tumbling unit must be eliminated. The filament die guides the impregnated rovings, which are wound on the rotating mandrel.

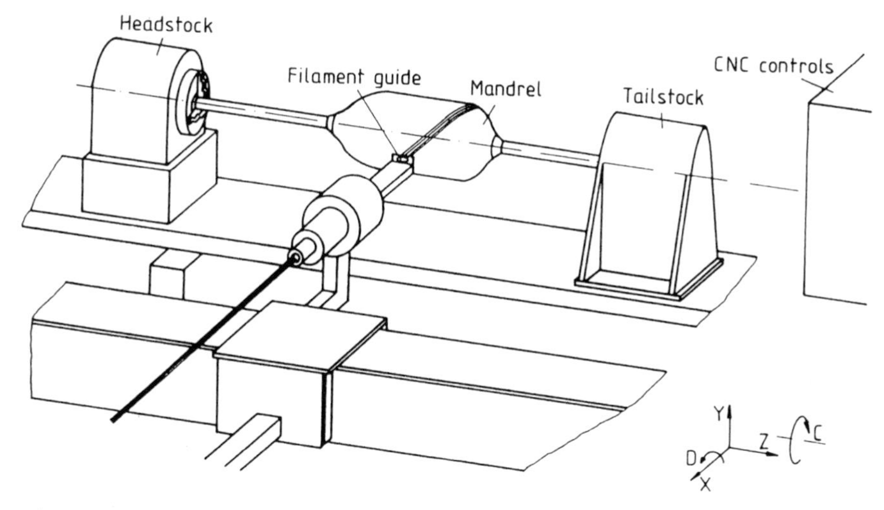

Figure 6.6.7 Lathe-type winding machine with mandrel.

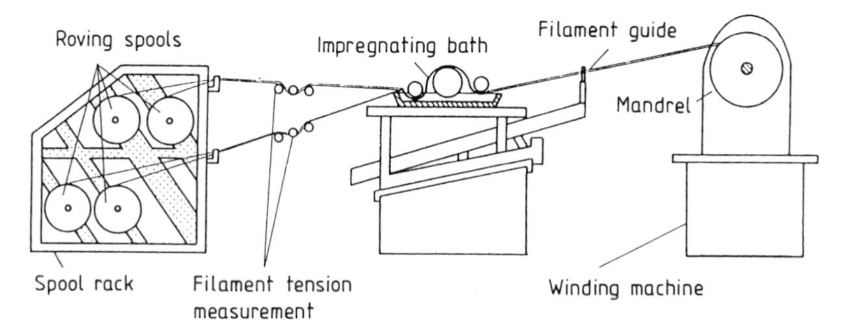

Figure 6.6.8 Winding machine with spool rack and impregnating bath.

The control system regulates the motion of the filament die so that one fiber orientation angle is maintained on the surface of the mandrel. Maintaining an optimum fiber orientation angle, especially on mandrels with complicated shapes, is limited by the slip of the rovings on the mandrel. A higher winding speed or greater number of rovings is employed to produce more rapidly and cost effectively. When winding with several rovings, they are fed through a wide filament die, or a so-called *circular filament* die is employed (Fig. 6.6.9). Another variation involves the winding of tapes, but they tend to bulge at the reversal points.

6.6.2.5 *Pultrusion*

Pultrusion is employed to produce fiber-reinforced continuous profiles. The basic arrangement of such a system is shown in Fig. 6.6.10. The previously dried fiber rovings are impregnated in a resin bath and then formed into the desired profile shape in a heated die.

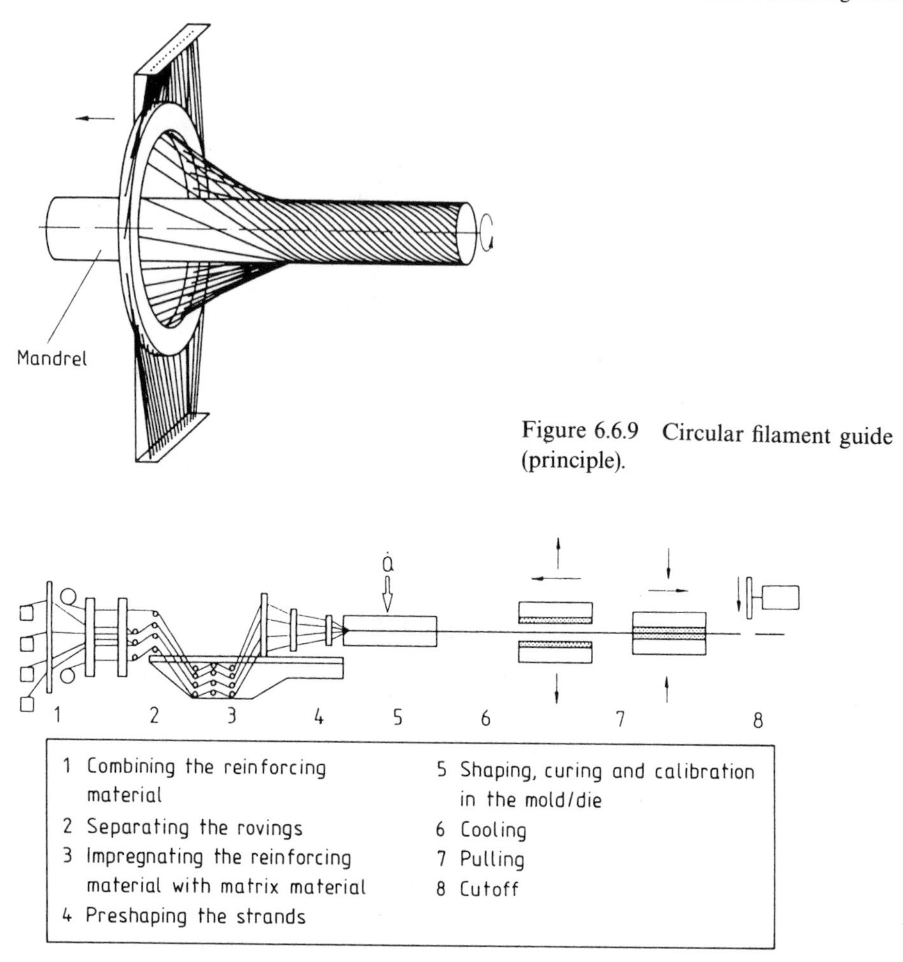

Figure 6.6.9 Circular filament guide (principle).

Figure 6.6.10 Process steps for pultrusion.

The heat applied cures the resin. Because of the high frictional forces that occur, the die should be as short as possible. On the other hand, the resin must cure to such an extent that the profile produced retains its shape in the following takeoff equipment. Because of the high frictional forces, hydraulic takeoff systems are often employed.

6.6.2.6 Compression Molding of Fiber-Reinforced Plastics

Compression molding represents a suitable production method whenever the objective is to produce large, flat parts with above-average mechanical properties. The good mechanical properties result from the glass fibers embedded in the molding compound. The gentle processing of compression molding—in comparison to injection molding—hardly damages the reinforcing fibers so that they are present in the molded part at practically

Type Components	Standard (wt. %)	LS (wt. %)	LP (wt. %)
1. Polyester resin	37.1	27.8	25.54
2. Thermoplastic component	—	9.3	14.60
3. Reagent	0.4	0.4	0.36
4. Release agent	1.5	1.5	1.50
5. Polyethylene powder	2.0	2.0	—
6. Fillers	55.6	55.6	56.50
7. Pigment dispersion	3.0	3.0	—
8. Thickening agent	0.4	0.4	1.50
Total	100.0	100.0	100.00

Figure 6.6.11.

their original length of several centimeters. Cycle times during production range between approximately 20 s to several minutes.

Molding Compounds

Sheet Molding Compound (SMC)

At present, SMC is the molding compound with the greatest economic importance. This molding compound consists primarily of unsaturated, i.e., hardening, polyester resins, the curing agent (catalyst), mineral fillers, the glass fibers, and various processing aids and additives. These include pigments for coloring, zinc stearate as lubricant and mold release agent, and magnesium oxide to thicken the resin. Fig. 6.6.11 lists three typical formulations. The LS (low-shrink) and LP (low-profile) formulations are molding compounds that have been modified through the addition of thermoplastic particles to achieve better surface quality of the molded parts (e.g., eliminate waviness or sink marks). The thermoplastic component can compensate for some of the shrinkage of the resin during the curing reaction.

SMC is produced on so-called SMC-producing systems (Fig. 6.6.12). In these systems, the premixed and still thin resin is spread onto supporting webs. One of these webs is then brought under a cutter to which continuous glass fiber roving strands are fed from spools. Depending on the setting, the rovings are cut into pieces 12–50 mm long in the cutter and fall via gravity onto the coated supporting web. This results in a thick, uniform, nonwoven fabric of glass fibers in which the glass fibers are statistically oriented. By laying down uncut roving strands, it is also possible to produce molding compounds that exhibit a unidirectional reinforcement with quasi-continuous fibers (option in Fig. 6.6.12).

The nonwoven fabric thus formed is then thoroughly impregnated with resin by being covered with a second coated web. As a rule, the resin mat has a thickness of 2 to 3 mm and is wound into large coils. Following this actual production process is a multiday ripening, during which the magnesium oxide causes the originally thin resin to thicken. The SMC reaches the production facilities as a leathery and tacky, but not stringy, mat. Since many SMC processors produce their own molding compounds themselves, there is a variety of resin formulations and fiber structures that is almost impossible to describe in detail. The term SMC is thus a general designation for a group of many molding compounds that differ from one another in composition.

a)

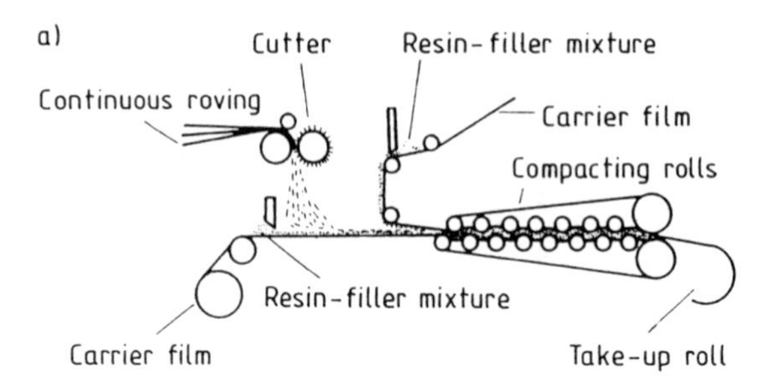

b)

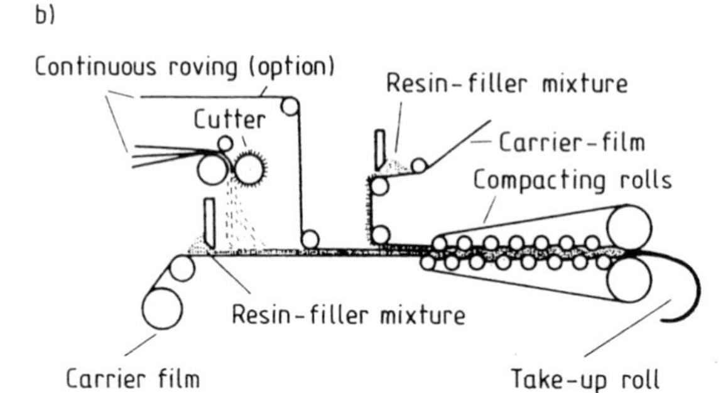

Figure 6.6.12 Principle of SMC-R and SMC-C/R production.

Typical applications for SMC include components for the electrical industry—e.g., wall panels for control cabinets, long lighting housings, etc.—and for the automotive sector. Spoilers and body components are produced from LP mats for utility vehicles, while hoods, rear deck lids, smaller louver panels for ventilation openings, and (with unidirectionally reinforced compounds) impact-resistant bumpers are produced for automobiles.

Glass-Mat-Reinforced Themoplastics

GMT, as SMC, is a general term covering a group of molding compounds. In contrast to SMC, however, GMT is available from only a few material suppliers as large sheets, since in this form it retains its shape (in contrast to SMC), is not tacky, and may be stored indefinitely. The various GMT grades differ from one another primarily in the structure of the glass fiber reinforcement. Through the use of different fiber lengths and glass fiber orientation, the processing and end-use properties can be varied over a wide range (from only moldable glass mats to flowable, dispersed fibers). Because of its good properties at a low price (Fig. 6.6.13), polypropylene (PP) is at present practically the only resin employed as matrix material.

Because of its exceptional impact toughness, which is retained even at low temperatures, in conjunction with a sufficiently high modulus of elasticity, GMT finds application for noise abatement beneath automobile engine compartments and in other regions exposed to the road. Other applications include large parts such as the highly stressed mechanical backbone in bucket seats and auditorium seating.

Preparing the Molding Compound

To prepare for the actual compression molding step, blanks must be cut from the sheets. With SMC this cutting can be done with simple knives, while with GMT impact shears

Advantages:
- low densities
- high modulus of elasticity
- high energy absorption
- high low-temperature impact strength
- similar values for unnotched and notched impact strength
- satisfactory heat distortion temperature [155°C]
- good chemical resistance
- production of large-area parts
- acceptable price

Disadvantages:
- undercuts hardly possible
- finishing required
- poor paintability (temperature resistance, surface, PP)
- alternative: textured, coated

Figure 6.6.13 Property profile of GMT-PP.

must be employed. The blanks are usually rectangular in order to keep the scrap from cutting at a minimum. The weight and wall thickness of the finished part are determined by the total number of blanks used for one molding cycle. The flow pattern in the compression mold is determined to a large extent by the shape of the blanks. As a rule, the individual blanks are placed in a stack prior to loading into the compression mold.

With SMC it is only necessary to remove the backing films, stack the blanks on one another, and load them into the compression mold. In contrast, the GMT must first be brought to a temperature above the melting point of the thermoplastic matrix in order to be moldable. This is accomplished in preheating stations that heat the individual blanks either via convection (recirculating air ovens), radiation (medium-wavelength infrared emitters), or conduction (contact heating).

After leaving the preheating station, the GMT blanks must be placed in a stack. This is a much more difficult operation than with SMC. In comparison to the leathery SMC, the preheated GMT blanks are even less rigid and extremely tacky because of the now molten matrix. In addition, the stack of blanks must be built up rapidly and placed in the compression mold in order to avoid cooling and loss of moldability.

Forming the Part via Compression Molding

As is the case with many other methods of plastics processing, compression molding also involves a coupling of thermal equalizing processes and flow processes during part formation. With SMC the cause of the thermal equalizing process is the temperature difference between the approximately 150°C compression mold and the room temperature molding compound. As the upper surface of the mold comes in contact with the stack of blanks, i.e., as part formation starts, heat transfer also takes place from this surface. Heating progresses further during part formation, making the SMC increasingly more fluid. When processing GMT, the compression molds are heated to approximately 80°C. The GMT thus starts to cool from the moment it is loaded into the mold.

The manner in which the molding compounds are prepared for compression molding means that the mass of molding compound metered into the mold is never really exact. For this

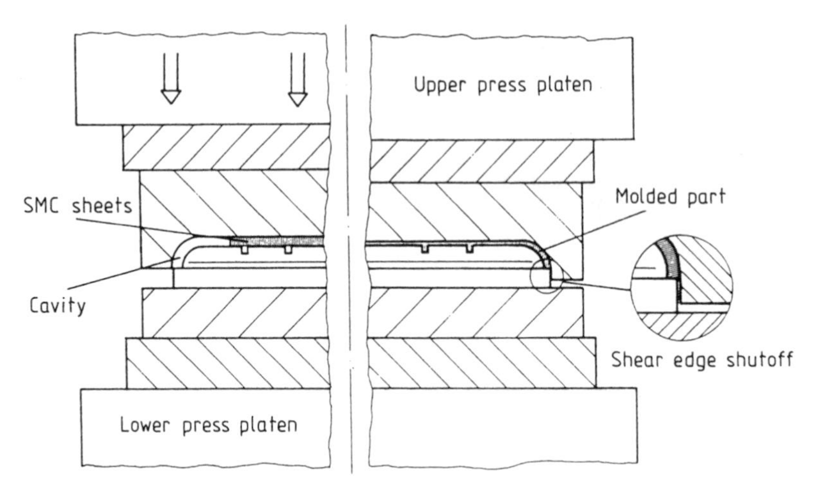

Figure 6.6.14 Compression molding of SMC.

reason, it is not possible to use a mold with a defined parting line. Instead, molds with so-called shutoff (shear) edges (Fig. 6.6.14) are employed.

As the mold closes, the shutoff edges of the two compression mold halves form a narrow gap in the direction of compression that prevents leakage of the molding compound. Slight variations in the amount of molding material thus translate only into a small difference in molded part thickness.

State-of-the-art presses (see example in Fig. 6.6.15) are often closed-loop controlled numerically to maintain parallelism in order to assure reproducible flow processes. Hydraulic presses with clamping forces up to several tens of thousands of kN are employed exclusively.

The flow process during molding of the stack of blanks is of decisive importance for the quality of compression molded parts, since the flow of the molding compound almost always gives rise to a preferred orientation of the originally statistically oriented glass fibers. Except for a few cases (where the flow-induced orientation accidentally coincides with the direction of load application in a highly-stressed mechanical component) such orientations are undesirable. Orientation leads to anisotropic shrinkage during curing and cooling and often results in distortion (warpage) of the molded part. The flow process can be influenced in a defined manner via the geometry of the stack of blanks and where it is placed in the mold.

Solidification of the Molding Compound

After molding, the SMC is at rest under a hydrostatic pressure in the closed mold. Contact with the hot compression mold leads to a further temperature increase, which finally activates the curing agent. Decomposition of the curing agent releases radicals that initiate curing, or cross-linking, of the unsaturated polyester resins. After passing through a viscosity minimum, the molding compound gradually becomes more viscous and finally turns into a rigid body (Fig. 6.6.16).

Because of the exothermic cross-linking reaction, the molding compound, and not the mold, releases additional heat as the degree of cure increases.

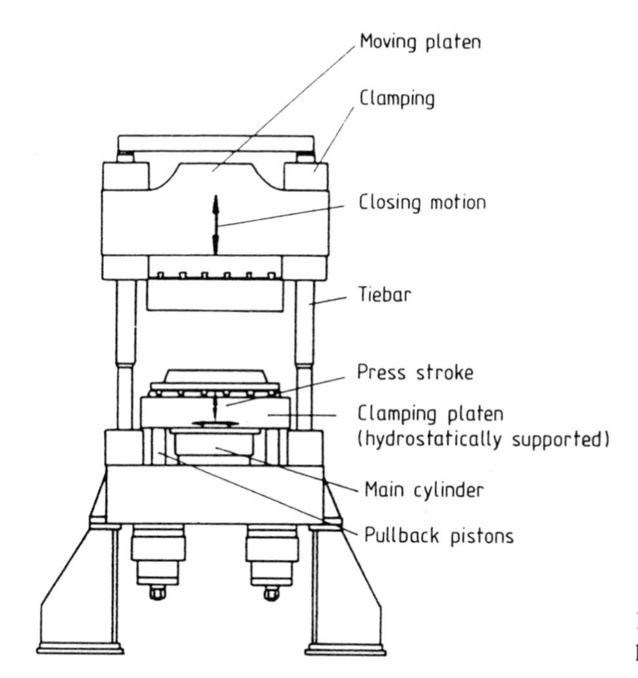

Figure 6.6.15 Short-stroke press with parallelism control.

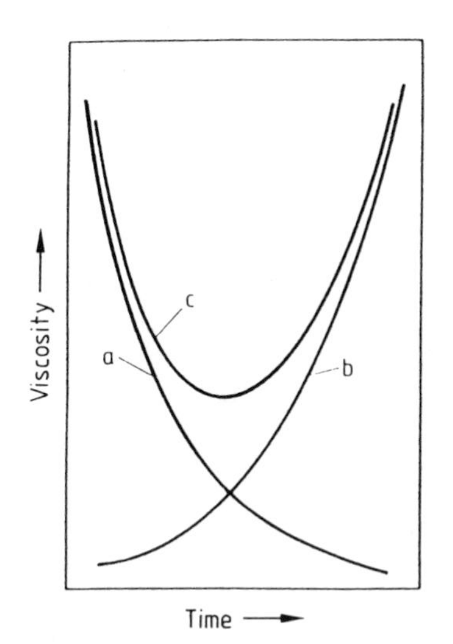

Figure 6.6.16 Qualitative viscosity curve during compression molding of SMC.

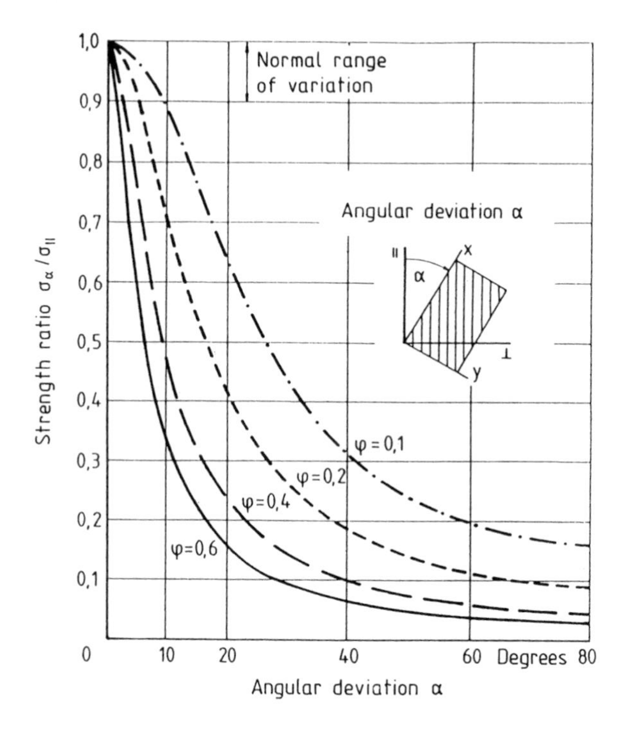

Figure 6.6.17 Effect of angular deviation on the strength of a UD ply.

GMT achieves the necessary dimensional stability simply through cooling and solidification of the thermoplastic matrix. This cooling begins immediately upon loading of the blanks into the mold. It is thus possible that the near-surface layers of the molding compound cool to such an extent during molding that they exhibit the characteristics of a solid. The GMT begins to solidify, starting at the surface of the mold. Molding and attainment of dimensional stability thus cannot be separated from one another, as is the case with SMC.

6.6.3 Part Design

In contrast to designing with metals, when designing parts of composites the part and the material must be "designed" together. The orientation of the fibers leads automatically to excellent properties in the fiber direction, but this is coupled, however, with a basic deficiency: The properties perpendicular to the fiber direction are considerably poorer. How significantly the strength depends on the angle of load application is shown in Fig. 6.6.17. For a fiber content of 60 vol.% ($\phi = 0.6$), an angular deviation of 5° from the principle load axis reduces the strength to only 60% of that possible in the fiber direction. The strength perpendicular to the fibers is only a small fraction of the tensile strength in the longitudinal direction. By building up layers of reinforcement in various directions, this basic deficiency can be eliminated. Unfortunately, however, this represents a compromise associated with considerable loss in material potential, as a comparison of such "quasi-isotropic" laminates with "unidirectional" laminates shows (Fig. 6.6.1). For instance, the modulus of elasticity decreases to about 1/3, while the strength drops to approximately 1/10 of the value for unidirectional laminates.

In view of this, the designer of fiber-reinforced composite parts has not only the often cited opportunity, but also the unavoidable obligation, to utilize the material properties to achieve a specific objective.

6.7 Calendering

Calender lines represent the largest and most expensive investments in the area of plastics processing. A calender consists of a set of precision rolls with which a prepared melt is formed. Calender lines are employed for the production of film and sheet (floor coverings) as well as for coating (e.g., coating of tire cords with rubber for the production of automobile tires). In plastics processing, calenders are used nowadays only for processing PVC compounds into film and floor coverings. Here it offers greater profitability than extrusion due to the very high output and savings in expensive stabilizers, since, in contrast to extrusion systems (blowing head, sheet die), the residence time of the resin in heating zones is much shorter (addition of 2% stabilizers corresponds to 50% of the PVC resin price!). In addition to the high output, the obvious advantages of a calender line include the easy access and possibility for rapid material changes (cleaning when changing batches, etc.).

For *rigid PVC* (without plasticizers) there are basically two different methods: the *Luvitherm process* and the *high-temperature (HT) process.* With the less commonly employed Luvitherm process for adhesive tapes, for instance, the PVC powder is merely sintered together on the calender (160°C). The tape melts and becomes highly transparent only upon passing over a heated roll (240°C) downstream from the calender.

Here we will consider in greater detail only the HT process that is used widely for PVC film. Flexible PVC calenders are not very much different in design (Fig. 6.7.1).

The primary components of a calender line for the HT process are shown in Fig. 6.7.1. These include:

- the weighing and premixing station (not shown),
- the plasticating system,
- the calender for forming and calibrating,
- the chill rolls,
- the winder,

and the equipment between these components for any stretching, embossing, or other processing.

In contrast to extrusion, plasticating, and shaping are separated. The individual stations are described in more detail in the following.

Weighing and Premixing Station

These functions are accomplished fully automatically today. The principle is similar to that employed in other large weighing systems for bulk goods.

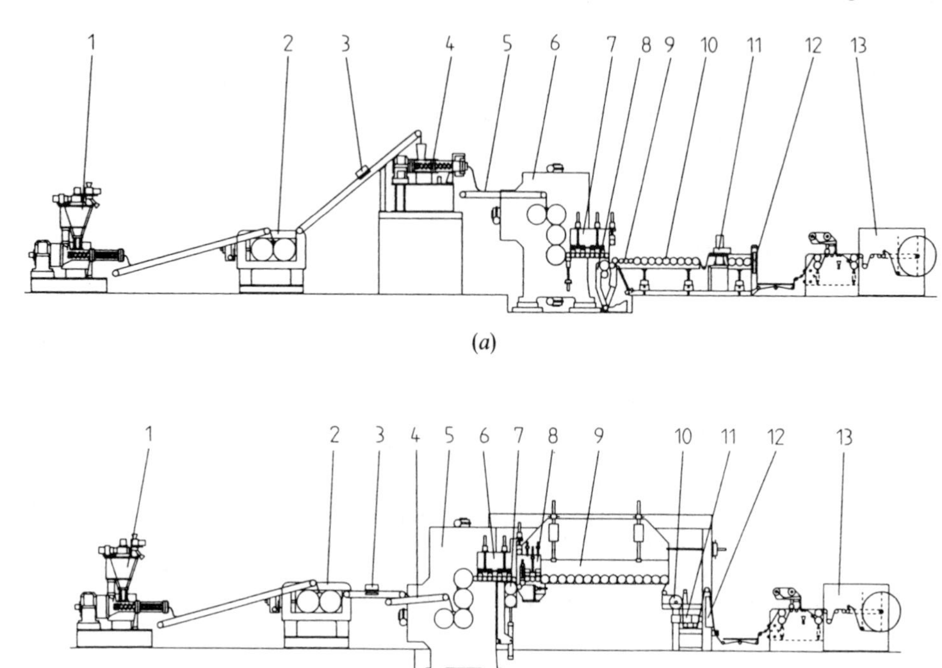

Figure 6.7.1 (a) Illustration of a calender line for production of rigid PVC film by means of the high-temperature process (Hermann Berstorff Maschinenbau GmbH, Hannover) 1 roller extruder, 2 roll mill, 3 metal detector, 4 strip feed (pivoting), 5 four-roll L-type calender, 6 take-off rolls, 7 embosser, 8 take-off rolls, 9 cooler, 10 vacuum roll, 11 thickness gauge, 12 illuminated control area (screen), 13 double winder; (b) Illustration of a system for producing flexible PVC film (Hermann Berstorff Maschinenbau GmbH, Hannover) 1 roller extruder, 2 roll mill, 3 metal detector, 4 strainer, 5 strip feed, 6 four-roll F-type calender, 7 take-off rolls, 8 embosser, 9 take-off rolls, 10 cooler, 11 thickness gauge, 12 illuminated control area (screen), 13 double winder.

Plasticating (Fig. 6.7.1a, Stations 1 and 2, as well as Fig. 6.7.1b, Stations 1, 2, 3, and 4)

After mixing the PVC powder with the necessary additives (plasticizers, stabilizers, lubricants, etc.), the mass is initially heated and mixed in batches in a *vortex mixer* (see Fig. 5.2.4) or, for high filler contents (floor coverings), in an *internal mixer*. A screw kneader, internal mixer, or planetary roller extruder (see Fig. 5.2.5a) and/or a roll mill follows the vortex mixer for plastication. The latter serves as a buffer station and forms a so-called rough sheet (sheet on the rolls), from which melt is usually obtained by cutting a strip at one end and feeding it to the first calender nip via the conveyor belt (Fig. 6.7.1a). To produce flexible PVC film (Fig. 6.7.1b), an additional strainer extruder (4), which further homogenizes the material and filters out any impurities, is incorporated prior to the first roll nip.

Forming Calender (Fig. 6.7.1a, Station 5, or Fig. 6.7.1b, Station 6)

The first roll nip is fed with the strip from the roll mill or strainer extruder via a feeding device. There the incoming strip is rolled to form the rough sheet, which does not yet have good surface quality. It then leaves the first calender nip and enters the second calender nip, where a "running vortex" forms (Fig. 6.7.2). Such vortices also form in the gap between the following pair of rolls. The vortices form because the resin mass builds up ahead of the narrow gap between the rolls. The rearward-flowing material in the vortex experiences pressure from the skin and the rolls. This material exhibits laminar flow, i.e., streamline after streamline of the incoming sheet is directed backwards. In this way, the surface of the sheet is "planed off" in a manner of speaking. As they flow backwards, the streamlines that previously formed the sheet surface flow into the central "kneading vortex" (see Fig. 6.7.2).

Slightly more than half of the incoming strip passes through the roll nip (thickness h_A) unaffected. The material drawn into the vortex is forced to the side (Fig. 6.7.3).

Such vortices are present in all roll nips. It can be seen from the figures that it is always the lower side of the incoming strip that passes through the roll nip unchanged, while streamline after streamline is stripped from the upper side and forced backwards. In this way, a new surface is formed at every nip. The quality of the sheet/film is thus determined by formation of the vortices. In order to achieve satisfactory quality, at least three nips are necessary: the first to form the rough sheet, the second and third to form each of the two surfaces anew. Each additional nip means a further improvement in quality. The sheet is always carried along by the hotter roll, that is, by the roll that is heated more or turns faster (frictional heat). Usually the calender rolls are arranged in the shape of an "F" or "L" (Fig. 6.7.4). The rolls are generally fabricated from a chill casting with a mirror finish. The diameter can be up to about 900 mm and the width up to 4500 mm. The surface quality and concentricity must satisfy extremely high requirements of $<2.5\ \mu m$. The roll nips are individually adjustable. The optimum film thickness lies between 200 and 250 μm. Thinner films are then produced by stretching at a downstream station.

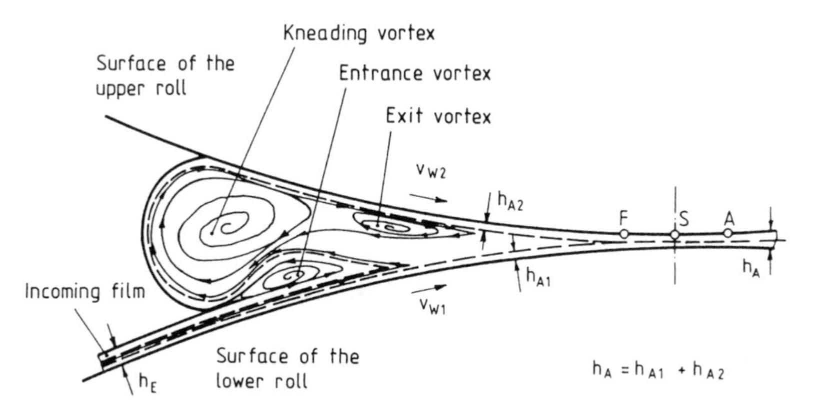

Figure 6.7.2 Schematic illustration of the flow conditions in the kneading vortes during calendering [according to Menges (1979)]; F = location where flows come together, S = narrowest part of nip, A = release location, $h_{A,E}$ = thickness of web, v_W = surface speed of roll, $v_W2 > v_W1$ (approximately 10%).

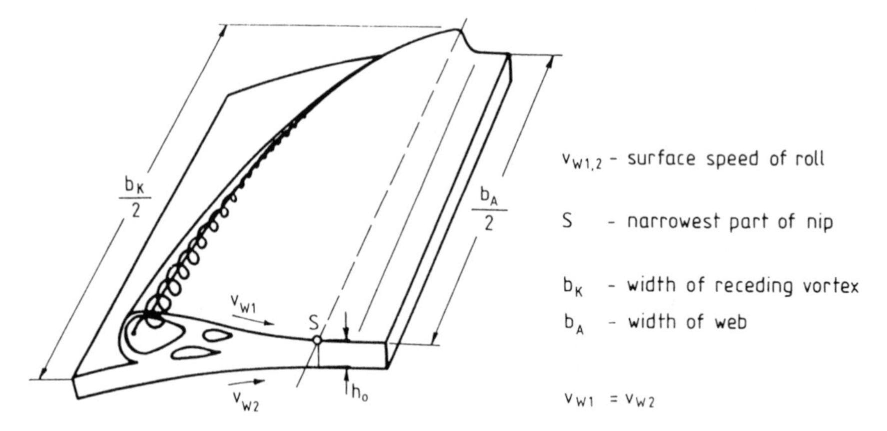

Figure 6.7.3 Helical motion of a material particle in the kneading vortex in the transverse direction; $v_W 1$ = surface speed of the upper roll, $v_W 2$ = surface speed of the lower roll, h_o = height of the narrowest spot in the nip, h_A = thickness of the calendered film, b_A = web width after leaving the nip, b_K = width of the kneading vortex, S = narrowest part of nip.

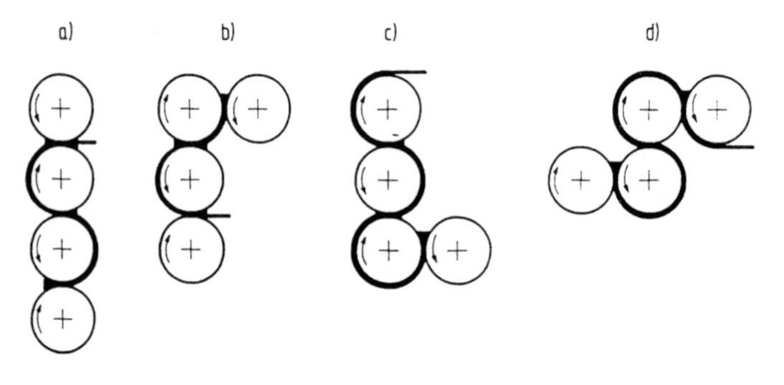

Figure 6.7.4 Arrangement of calender rolls.

Hot water is best suited for heating, since a temperature of ± 1°C must be assured over the entire width of the roll. When processing rigid PVC, the temperature increases from roll to roll from about 180 to over 200°C. To facilitate release, the last roll is run somewhat cooler.

Maintaining a uniform gap over the entire width of the roll is a difficult mechanical engineering problem. To compensate for deflection, there are basically the possibilities shown in Fig. 6.7.5. The simplest solution utilizes a convex, or crowned, roll. In this case, the rolls initially are ground in such a way as to just compensate for the deflection. Depending on the material viscosity, however, this may no longer be possible. Other possibilities include roll bending and crossing over to compensate for the deflection.

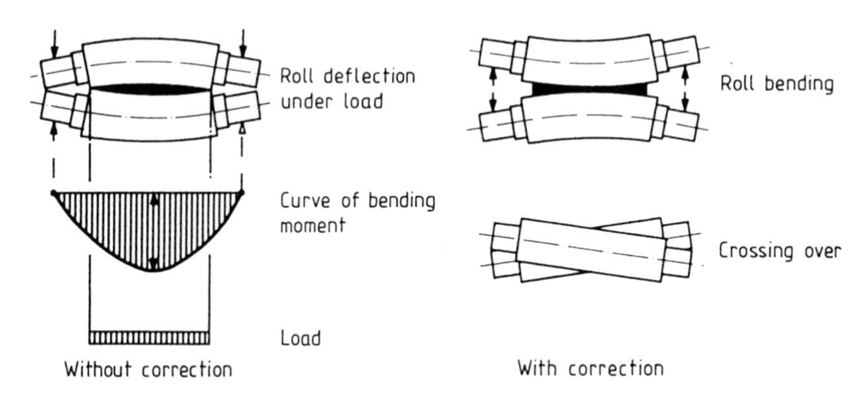

Figure 6.7.5 Possible corrections for roll deflection.

Chill Rolls (see Fig. 6.7.1, Station 9)

The film is pulled off the last calender roll by a small-diameter take-off roll and then passes over chill rolls. These are large-diameter, driven cooled rolls, for instance, for flexible PVC floor covering, which requires additional cooling because of its relative thickness. In contrast, film cools by radiation and convection when passing over temperature-controlled guide rolls.

Take-Off Rolls

The film or sheet is removed with the aid of take-off rolls having a large wrap angle in order to assure that the sheet is always under constant tension, thereby eliminating any effects of the fluctuating pull of the reel on the film speed in the calender line.

Winder (see Fig. 6.7.1a, Station 13)

The film is wound by winders with a fast changing system. These winders can transfer the film automatically without interrupting production as soon as one reel is filled. The speeds are about 40 m/min for rigid PVC, about 60 m/min for flexible PVC, and about 180 m/min for film stretched on-line.

Secondary Processing (Downstream Equipment)

Often secondary processing steps such as embossing and, very often, stretching stations are incorporated into the calender line. Stretching of film is used to achieve an increase in strength and (or) a reduction in thickness. A less practical solution would be to first reheat the sheet and then stretch it. With the stretching method most commonly employed today, the film passes over many small rolls in succession with the last roll driven faster than the first, thereby stretching the film. The rolls in between are free-running. All of the rolls are maintained at the stretching temperature.

6.8 Casting

Characteristic of this processing method is that the resins being processed flow solely under the influence of gravity or the influence of centrifugal forces (*centrifugal casting*). Parts are produced without the use of pressure in open or closed molds. Film is cast onto drums (wheels) or as continuous sheets. This processing method is employed for both thermosetting and thermoplastic resins.

A prerequisite for use of this method is that the resin be in a flowable condition. The resins can be in the form of melts, solutions, dispersions, powder (thermoplastics), or as curing *casting resins* (thermosets). After *forming* of the molded part, the resin solidifies via cooling (thermoplastics) or the cross-linking reaction (thermosets).

Advantages of this processing method include:

- lower expenses for molds, since the largely pressureless processing means that the mechanical loads on the mold are low so that the molds need not be built as ruggedly;
- simpler machine requirements for molded part production;
- more economical production of small quantities;
- the particularly homogeneous and isotropic structure of cast parts.

The following disadvantages, however, must also be taken into consideration:

- with thermosetting resins, air is often incorporated during mixing and under certain circumstances can no longer escape because of the increase in viscosity as a result of the curing reaction; casting of bubble-free parts thus requires that the mold be evacuated;
- the cross-linking reaction of casting resins is exothermic in many cases so that, in the absence of adequate heat removal, excessive heating of the interior of the molded part, and thus cracking, may occur because reaction takes place too rapidly;
- with very thick-walled parts, the curing times are very long because of the heat of reaction, which, in turn, leads to longer cycle times and the need for several molds to achieve economical production, so that the advantage of lower mold costs is offset in part;
- when processing solvent-containing resins, it is necessary to recover the solvent.

The following variations are distinguished among the casting processes:

- casting proper,
- centrifugal casting (rotational casting),
- film casting,
- encapsulating,
- impregnating.

Casting

Casting of thermosetting resins involves the following sequence of steps:

- compounding,
- mixing the resins,

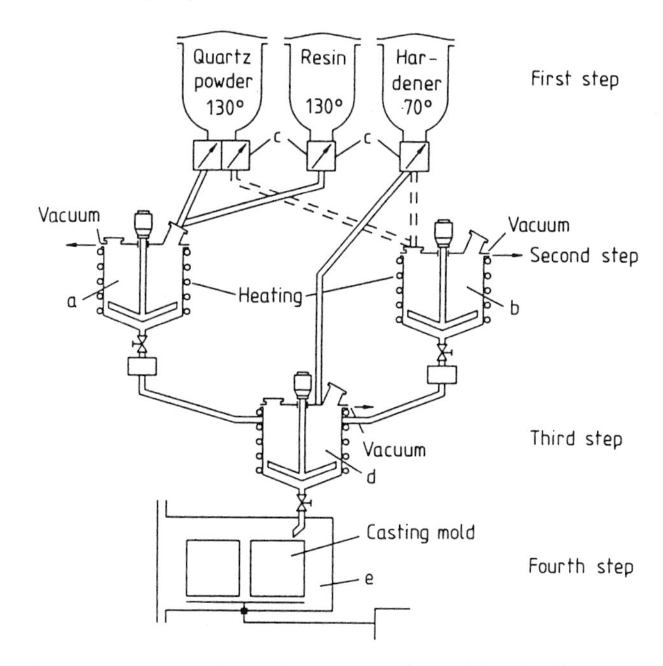

Figure 6.8.1
Casting system.

- casting,
- curing in the mold,
- part removal and possibly post-curing of the molded part in an oven, and finishing.

Of the thermosetting materials, epoxy resins are preferred for casting. These resins are low-pressure resins, i.e., they can be processed with little or no pressure, since there are no by-products produced during the cross-linking.

Fig. 6.8.1 shows a four-stage vacuum casting system for molding parts of filled, heat-curing epoxy resins. All containers are maintained at an elevated temperature to lower the viscosity of the individual components.

Compounding (Stages 1 and 2)

Usually the fillers are incorporated into the resin component, which has the greater volume. For high-quality parts that are not permitted to contain any entrapped air, the fillers are mixed with the resin under vacuum in the second stage. This is necessary for electrical components such as insulators, where gas inclusions considerably reduce the insulating capacity.

Mixing (Stage 3)

To avoid the inclusion of gas, addition and mixing of the hardener also takes place under vacuum. The temperature must be controlled extremely accurately at this stage so that the reaction does not start prematurely, thereby reducing the viscosity.

Casting (Stage 4)

Casting of the mixture can take place with or without vacuum. Any solid material capable of withstanding the curing temperatures to be expected is suitable as material for the casting molds. Since epoxy resins are very adhesive, a mold release must always be applied to the mold surfaces prior to casting of the resin.

Curing

With heat-curing resins, temperature control of the oven must take into consideration the change in temperature as a result of the cross-linking reaction of the resin.

Part Removal, Post-Curing, and Finishing

As a rule, the part is removed manually. To achieve a high throughput with the casting molds, the shortest possible cycle time is selected and the parts are post-cured in an oven. Finishing to remove the sprue and risers is usually required as the final step. Because of their high dimensional stability under heat and poor flammability, cast parts find application primarily as electrical components.

Powder sintering is a sintering process used to produce parts from thermoplastic powders (Fig. 6.8.2). With this process, the cavity is filled completely with the powder and the mold is heated in an oven. While in the oven, the powder in the vicinity of the cavity surface

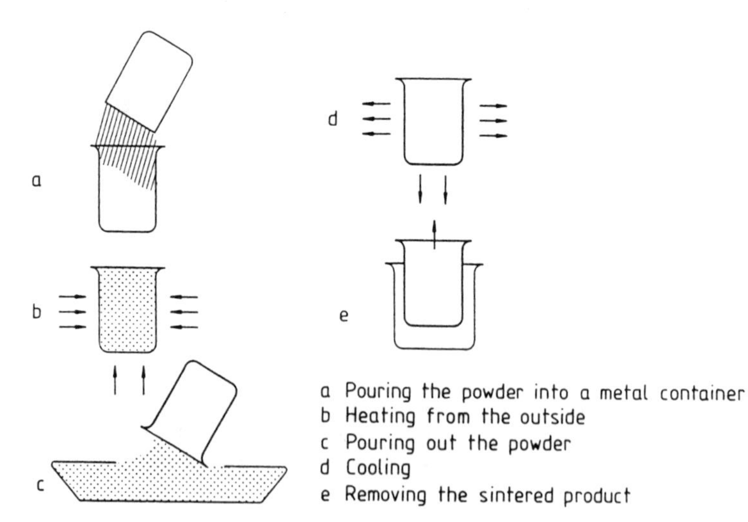

Figure 6.8.2
Powder sintering.

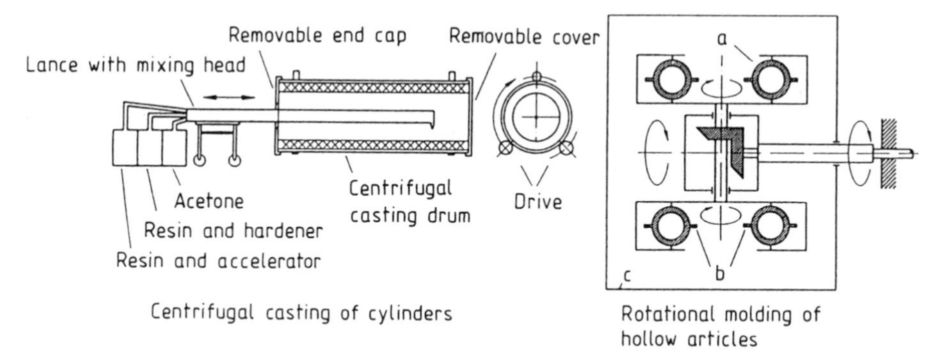

Figure 6.8.3 Centrifugal casting/rotational molding.

starts to melt. In the next step, the unmelted powder is poured out and the mold is heated once again in order to smooth the inner surface of the molded part. After subsequent cooling and solidification, the part can be removed from the mold.

This process is employed primarily for the production of large hollow bodies such as containers, especially when the quantity of parts to be produced is too low to justify blow molding or injection molding. Since the process is a quasi-pressureless one, lightweight molds of simple mechanical design can be employed.

Centrifugal Casting (Rotational Casting/Molding)

Plastic parts can also be produced in rotating molds in a manner analogous to the centrifugal casting used to produce bimetallic barrels (Fig. 6.8.3). To produce pipes only one axis is required.

The starting material can be either a casting resin or a thermoplastic powder that melts in the rotating heated mold and is subsequently distributed by the centrifugal force. In this way it is possible, for instance, to manufacture large-diameter polyethylene pipe that cannot be produced by extrusion.

Film Casting

In film casting the casting compound is distributed uniformly on a horizontal surface. The compound subsequently solidifies on this carrier. The continuous mode of operation with an endless belt (belt casting machine) or rotating drum (drum casting machine) as a carrier is illustrated in Fig. 6.8.4. Advantages of cast film include the uniform, orientation-free structure and higher degree of crystallization that can sometimes be achieved.

Encapsulation

Many electrical components are protected from environmental factors by being encapsulated in plastics. As a rule, casting resins are employed for this purpose. The process technology does not differ from conventional casting, except that an insert is involved.

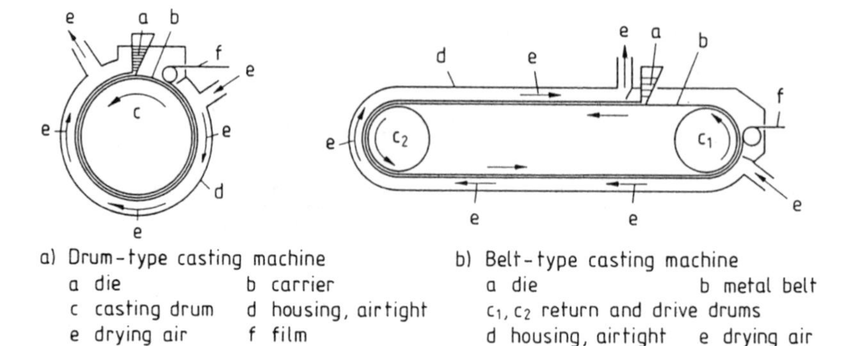

Figure 6.8.4 Film casting.

Impregnation

Impregnation is the process of soaking a flat, porous material (e.g., paper, woven and nonwoven fabrics, mats) with resins, solutions, or dispersions. The objective of this treatment is to improve the mechanical properties (paper, textiles) or to reduce the effect of moisture on the properties of the material to be impregnated.

Bibliography for Chapter 6.1

Michaeli, W.: Extrusionswerkzeuge für Kunststoffe und Kautschuk, Bauarten, Gestaltung, Berechnung, 2nd ed., Carl Hanser Verlag, München, Wien, 1991
Schenkel, G.: Kunststoff-Extrudertechnik, Carl Hanser Verlag, München, Wien, 1963
Wortberg, J.: Werkzeugauslegung für die Ein- und Mehrschichtextrusion, Doctor theses at RWTH Aachen, 1978
Hensen, F., Knappe, W. Potente, H.: Handbuch der Kunststoff-Extrusionstechnik, 2 volumes, Carl Hanser Verlag, München, Wien, 1986, 1989
Menges, G.: Einführung in die Kunststoffverarbeitung, 2nd ed., Carl Hanser Verlag, München, Wien, 1979
Predöhl: Technologie extrudierter Kunststoffolien, VDI-Verlag GmbH, 1979
Domininghaus, H.: Fortschrittliche Extrudertechnik, VDI-Verlag, 1970
Feistkorn, W.: Auslegung und Optimierung von Plastifizierextrudern, Doctor theses at RWTH Aachen, 1985
Potente, H.: Modellgesetze für Ein- und Zweischneckenmaschinen, Carl Hanser Verlag, München, Wien, 1981

Bibliography for Chapter 6.2

Kulik, M.: Ein Beitrag zur Prozeßanalyse des kontinuierlichen Extrusionsblasformens, Doctor theses at RWTH Aachen, 1974
N. N.: Qualität sichern im Blasformbetrieb, VDI-Verlag, Düsseldorf, 1979
N. N.: Extrusionsblasformen, VDI-Verlag, Düsseldorf, 1979
Hüsgen, U.: Thermische und rheologische Berechnungen im Bereich Blasformen, Doctor theses at RWTH Aachen, 1988

N. N.: Spritzblasen, VDI-Verlag, Düsseldorf, 1976
Rosato, D.V.: Blow Molding Handbook, Carl Hanser Verlag, München, Wien, New York, 1989
N. N.: Technologien des Blasformens, VDI-Verlag, Düsseldorf, 1977
Holzmann, R., Lesch, H.D.: Blasformanlagen – aus Sicht des Verarbeiters in: Automatisierung in der Kunststoff-Verarbeitung, Carl Hanser Verlag, München, Wien, 1986
Junk, P.B.: Betrachtungen zum Schmelzeverhalten beim kontinuierlichen Extrusionsblasformen, Doctor theses at RWTH Aachen, 1978
Esser, K.: Untersuchungen zur Prozeßführung beim Streckblasformen von PP, Doctor theses at RWTH Aachen, 1985

Bibliography for Chapter 6.3

Menges, G., Mohren, P.: Anleitung zum Bau von Spritzgießwerkzeugen, 3rd ed., Carl Hanser Verlag, München, Wien, 1991
Lanvers, A., Breuer, P.: Spritzgießen, Lecture notes IKV, RWTH Aachen, 1990
Menges, G.: Einführung in die Kunststoffverarbeitung, 2nd ed., Carl Hanser Verlag, München, Wien, 1979
Mink, G.: Grundzüge der Spritzgußtechnik Zechner & Hüthig Verlag, 5th ed., 1990
Menges, G., Porath, U., Thim, J., Zielinsky, J.: Lernprogramm Spritzgießen, Carl Hanser Verlag, München, Wien, 1980
Elbe, W.: Der Plastifiziervorgang – Verfahrenstechnische Forderungen und Erkenntnisse in: Spritzgießtechnik, VDI-Verlag GmbH, Düsseldorf, 1980
Thoma, H.: Vollhydraulik- oder Kniehebelsystem Kunststoffberater 1981, issue 2, p. 821–825
Wiegand, H.-G.: Prozeßautomatisierung beim Extrudieren und Spritzgießen von Kunststoffen, Carl Hanser Verlag, München, Wien, 1979
Sauerbruch, E.: Aufbau und Funktion einer neuartigen Spritzgießmaschine
Tadmor, Z., Gogos, C.G.: Principles of Polymer Processing, J. Wiley & Sons, New York, 1979
Stoeckhert, K.: Werkzeugbau für die Kunststoff-Verarbeitung, 3rd ed., Carl Hanser Verlag, München, Wien, 1979
Gastrow, H.: Der Spritzgieß-Werkzeugbau in 100 Beispielen, Carl Hanser Verlag, München, Wien, 1990
Sarholz, R., Beese, U., Hengesbach, H.A., Wübcke, G.: Spritzgießen, Verfahrensablauf, Verfahrensparameter, Prozeßführung, Carl Hanser Verlag, München, Wien, 1979
Robers, Th.: Spritzgießmaschinentechnik, Lecture note IKV, RWTH Aachen, 1990
Johannaber, F.: Spritzgießmaschine in: Kunststoffmaschinenführer, 3rd ed., Carl Hanser Verlag, München, Wien, 1979

Bibliography for Chapter 6.4

Becker, W., Braun, D.: Kunststoffhandbuch, volume 10: Duroplaste, edited by W. Woebcken, Carl Hanser Verlag, München, Wien, 1988
Burns, R.: Polyester Molding Compounds, Marcel Dekker, Inc., New York, 1982
Meyer, R.W.: Handbook of Polyester Molding Compounds and Molding Technology, Chapman and Hall, New York, London, 1987
Menges, G.: Werkstoffkunde Kunststoffe, 3rd ed., Carl Hanser Verlag, München, Wien, 1990
Schönthaler, W.: Verarbeiten härtbarer Formmassen, VDI-Taschenbuch T40, VDI-Verlag GmbH, Düsseldorf, 1973
Saechtling, H.: Kunststofftaschenbuch, 25th ed., Carl Hanser Verlag, München, Wien, 1992

Bibliography for Chapter 6.5

Menges, G.: Einführung in die Kunststoffverarbeitung, 2nd ed., Carl Hanser Verlag, München, Wien, 1979
Müller, H.: RIM-Technologie – Beitrag zur Verbesserung und Sicherung der Fertigung technisch hochwertiger Formteile, Doctor theses at RWTH Aachen, 1985
Maier, U.: Auslegung von Werkzeugen zur Fertigung von PUR-Formteilen nach dem RIM-Verfahren, Doctor theses at RWTH Aachen, 1987
Mrotzek, W.: Gegenstrominjektionsvermischung niedrigviskoser Reaktionsharze, Doctor theses at RWTH Aachen, 1982
Grasse, H.: Bausteine für Anlage zur Herstellung von RRIM-Formteilen, Kunststoffe; 76/1986, no. 12, p. 1181–1184
Braun, H.-J., Eyerer, P.: PUR-RIM und RRIM-Technologie: Fortschritte und Wirtschaftlichkeit Kunststoffe; 78 (1988) 10, p. 991 ff
Begemann, M., Maier, U., Müller, H., Pierkes, L.: RIM–die Formteilqualität verbessern (Teil 1): Homogenisierung von Mehrkomponentensystemen Plastverarbeiter; volume 37, no. 6/1986, p. 52 ff
Lee, L.J.: Reaction Injection Molding Comprehensive Polymer Science, Volume 7, S. 379–426
N. N.: Reaction Injection Molding Encyclopedia of Polymer Science and Engineering (1989) 14, p. 72–100, John Wiley & Sons, New York, Brisbane, Toronto, 1989
Becker, W., Braun, D.: Kunststoffhandbuch, volume 7: Polyurethane, edited by G. Oertel, Carl Hanser Verlag, München, Wien, 1983
Macosko, C.W.: RIM–Fundamentals of Reaction Injection Molding, Carl Hanser Verlag, München, Wien, New York, 1989

Bibliography for Chapter 6.6

Fitzer, E.: Wehrtechnisches Symposium am Wehrwissenschaftlichen Institut für Materialuntersuchung 1984, Erding, p. 3.01–3.39
Hillemeier, K.: Aramidfaser in: Kohlenstoffaser- und Aramidfaserverstärkte Kunststoffe, VDI-Verlag, 1977, p. 29–44
Lang, R.W., Stutz, M., Heim, M., Nissen, D.: Polymere Hochleistungs-Faserverbundwerkstoffe "Makromolekulare Chemie" der GDCh in Bad Nauheim, 14./15. April 1986, Tagungsumdruck
Niederstadt, G.: Übersicht über die Vielfalt der Verbundkonstruktionen, VDI-Bildungswerk
N. N.: Du Pont Information on Kevlar
Öry, H., Reimerdes, H.G.: Faserverbundwerkstoffe, Lecture notes Faserverbundwerkstoffe, RWTH Aachen, 1987
Rebenfeld, L.: The interface between fibres and resins in composites, 25. Internationale Chemiefasertagung 1986, Dornbirn
Michaeli, W., Wegener, M.: Einführung in die Technologie der Faserverbundwerkstoffe, Carl Hanser Verlag, München, Wien, 1989
Reinhard, T.J., et al.: Composite, Engineered Materials Handbook, Vol. 1, ASM International, 1987
Schwarz, O.: Glasfaserverstärkte Kunststoffe, kurz und bündig, Vogel-Verlag, 1975
Menges, G.: Einführung in die Kunststoffverarbeitung, 2nd ed., Carl Hanser Verlag, München, Wien, 1979
N. N.: Formteile und Profile aus GFK für den industriellen Einsatz, Editor: Verein Deutscher Ingenieure, VDI-Gesellschaft Kunststofftechnik, Düsseldorf 1980
Martin, J.D., Sumerak, J.E.: Pultrusion, Engineered Materials Handbook, Vol. 1, Composites, ASM International, Ohio, 1987, p. 533–543

Meyer, R.W.: Handbook of Pultrusion Technology, Chapman and Hall, New York, London, 1985
Sumerak, J.E.: Understanding Pultrusion Process Variables for the First Time, SPI, 40th annual conference, February 1985, Atlanta, USA

Bibliography for Chapter 6.7

Menges, G.: Einführung in die Kunststoffverarbeitung, 2nd ed., Carl Hanser Verlag, München, Wien, 1979
Kopsch, H.: Kalandertechnik, Herstellung von Folien und Tafeln nach dem Walzwerkverfahren, Carl Hanser Verlag, München, Wien, 1978
Breuer, H.: in: Hensen, Knappe, Potente, Handbuch der Kunststoff-Extrusionstechnik volume II: Extrusionsanlagen, Carl Hanser Verlag, München, Wien, 1986

Bibliography for Chapter 6.8

Menges, G.: Einführung in die Kunststoffverarbeitung, 2nd ed., Carl Hanser Verlag, München, Wien, 1979
Rost, A.: Verarbeitungstechnik der Epoxid-Harze, Carl Hanser Verlag, München, Wien, 1963
Stierli, R.: Einführung in die Gießharztechnik in: Kunststoff-Handbuch Volume 10: Duroplaste, Carl Hanser Verlag, München, Wien, 1988

7 Secondary Processing Methods for Plastics

7.1 Thermoforming

The production of three-dimensional parts from plastic film or sheet through the action of heat and compressed air or vacuum has developed into a fully automated, industrial production process from its beginnings as a largely manual trade in the early 1950s. In principle, this forming method can utilize any of several processing techniques. When forming thermoplastic materials, the energy required for deformation is generally introduced by means of vacuum and/or compressed air in conjunction with a mechanical assist. Processing takes place in a region where the material exhibits rubberlike elasticity. In contrast, the primary processing methods of extrusion and injection molding take place in a temperature range where the materials are present as melts. A characteristic of forming methods is that the starting materials must first be brought to the forming temperature from room/storage temperature, for which reason the term *thermoforming* has become accepted in international usage as a general designation for all process variations.

The spectrum of products produced with this method ranges from packaging containers (quantities of up to 100,000/h) to large parts such as swimming pools measuring 8 m × 4 m × 1.5 m (quantities of 1.5/h). In addition, it is an important method for the production of automotive components (instrument panels, interior door panels).

Materials processed are predominantly amorphous plastics (PS, ABS, SAN, CAB, PMMA, PC, PVC) and semi-crystalline materials (PP, PE) as well as special multilayer film systems that economically combine the good properties of several plastics in a single product and—depending on the combination—exhibit barrier properties with regard to gas or liquid diffusion in the field of food packaging.

7.1.1 Machines

Basically, two systems have proven successful as industrial thermoforming machines today:

- single-station machines,
- multistation machines.

In the single-station machine (Fig. 7.1.1) heating and forming take place at the same station. Usually the heating system is moved into position above the clamped sheet and then retracted prior to forming; occasionally the clamped sheet is moved into the heating zone. The times required for the individual process steps of loading, heating, retraction of the heating system, preforming, final forming, cooling, and part removal add together to yield the overall cycle time.

In the multistation machine (Fig. 7.1.2), heating and forming take place in spatially separated stations. The slowest process step determines the overall cycle.

If the material is fed to the machine cut to size, the machine is called a sheet-fed machine. Loading and unloading can be accomplished either manually or automatically. The stock thickness can range between 0.1 and 12 mm. As a rule, single-station machines are fed sheets and are thus universal in their application. Simpler than loading from a stack of sheets is working from a roll of film on a roll-fed machine. Generally, thermoplastic film can be unwound from a roll up to a thickness of 2.5 mm. Roll lines find application in mass production, such as that for packaging containers or automotive interior panels. Fig. 7.1.2 shows a line with a punching station after the thermoforming station in an arrangement used when the quantities are not too high and the requirements for punching accuracy are not too stringent. With more stringent requirements for accuracy, punching also takes place in the mold, i.e., a combined thermoforming mold/punching die is employed, which assures uniform punching and reduces the so-called punching offset. Such molds are, however, more complex and generally have a shorter service life than do single-purpose thermoforming molds with a separate punching station. Nevertheless, complex mold arrangements and designs are justified to minimize scrap.

If, starting from a roll, the containers are formed, filled, and sealed with plastic-coated aluminum foil or with plastic film in one machine, the system is called a "form-fill-seal line" (FFS line).

One of the simplest and least expensive means of packaging small and light-weight parts and even machine components up to 50 kg in weight is skin packaging. The goods to be packaged are placed on a card coated with adhesive; the film is heated while at the same time the card with the goods is placed in the forming station, where it is forced against the

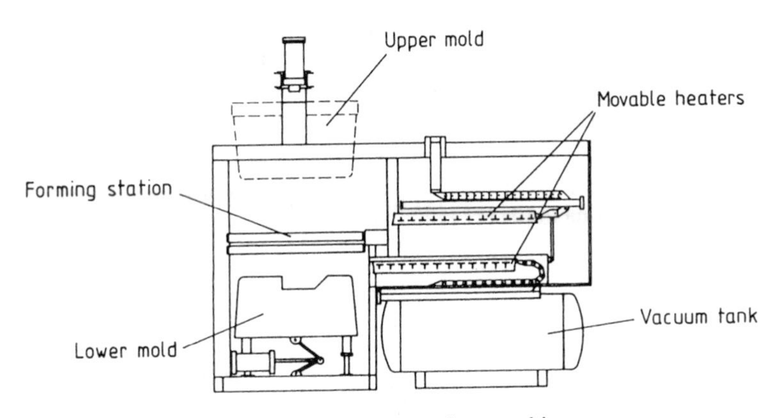

Figure 7.1.1 Components of a single-station machine.

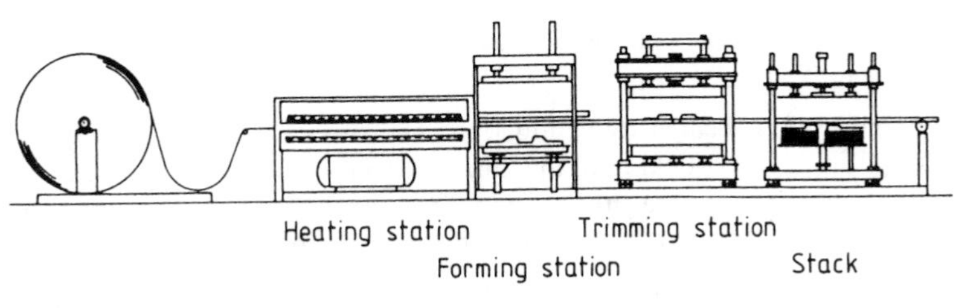

Figure 7.1.2 Components of a multistation machine.

heated film. By exhausting the entrapped air, the film is drawn tightly against the items and card. Skin packaging does not require a mold, since the items to be formed represent the mold over which the heated film is drawn with the aid of vacuum.

With so-called *blister packaging*, thin, transparent packaging pouches are filled with the items to be packaged and then heat-sealed with a card, as with skin packaging. The blister pouches can be produced on automatic thermoforming machines of various types. For small quantities, the pouches are filled manually, while automatic filling machines are employed for larger quantities. After automatic heat sealing with the aid of pressure and heat in the sealing station, the finished package is shipped.

7.1.2 Process Steps

Thermoforming of plastics can be divided into three basic process steps:

- the heating step,
- the change-over and forming step,
- the cooling step.

Heating Step

As the initial step for all thermoforming processes, the heating step is of decisive importance with regard to the subsequent quality of the thermoformed parts. In this process step, the material to be processed is heated from its initial temperature to the forming temperature. Since the thermoforming process takes place in the region of the elongation maximum, amorphous plastics are heated to temperatures above the softening point and semi-crystalline ones to temperatures around the crystallite melting point.

Since stock ranging in thickness from about 0.1 mm to about 12 mm is processed by means of thermoforming, various heating methods are employed:

- convection heating,
- contact heating,
- infrared heating.

All methods can be employed for either one or two sides. Except for when processing very thin films, two-sided heating is generally employed. *Convection heating* has the advantage of being independent of size. The surface temperature increases in the course of heating. The long heating times that result from the low thermal conductivity of plastics and the low heat transfer coefficients give rise to only slight temperature differences over the cross section of the material to be processed. Extremely thick plastic sheets are often heated slowly and gently in this manner.

With *contact heating*, the stock is heated by conduction. The surface temperature remains constant during heating. To prevent sticking of the film, the heating plates are covered with a nonstick coating. Heating plate temperatures around 200°C are employed today, and this method is used primarily to heat thin films in order to achieve short and uniform heating.

Because of its universality, *infrared heating* is the most commonly encountered heating method today. It has the great advantage that a portion of the energy can penetrate directly into the interior of the material. In this way it is possible to achieve higher heat fluxes in order to reach shorter heating times without thermally degrading the surface of the plastic. Ceramic and quartz emitters are the predominate radiation sources. Modern thermoforming systems are of modular design with a large number of infrared emitters. By varying the intensity of the individual emitters, the heating is matched to the requirements of the material and the part to be formed. Thermoforming systems are now being equipped increasingly often with infrared pyrometers that measure the stock temperature at a certain location on the surface and initiate the thermoforming cycle upon reaching the desired surface temperature. To control the heating process even more accurately, the surface temperature of the stock is measured subsequent to leaving the heating station, and the emitter intensities are controlled in accordance with a specified field of desired temperatures.

Change-Over and Forming Step

Thermoforming is a process involving only extension, which is performed in a continuous manner. The material to be formed is firmly held around its edges. Accordingly, the increase in surface area resulting from forming leads to a reduction in thickness.

The forming process begins upon completion of the heating step and ends once the film or sheet is in contact with the mold surface. For this process step the heating is retracted (single-station machine) or the stock is transferred from the heating station to the forming station (multistation machine).

Forming Methods

With all process variations of thermoforming, the stock, which has been heated to a rubberlike elastic condition, is rapidly subjected to biaxial stretching. This step generally lasts less than one-half of a second and thus can be assumed to be isothermal as long as free, or unrestricted, stretching is involved. The stock cools, however, immediately upon contact with the mold surface. As a result, the resistance to deformation increases spontaneously, especially with semi- crystalline thermoplastics. Mechanical aids, compressed air, and vacuum are employed to apply the necessary force.

When thermoforming, a basic distinction is made between *positive (male mold)* and *negative (female mold)* forming methods. A number of process variations derive from these two basic techniques, all of which strive to achieve a specific wall thickness distribution. The decision as to whether a positive or negative forming method is employed depends primarily on which side of the part must exhibit exact replication. Only the side in contact with the mold is dimensionally accurate; the other side will always exhibit variations, especially the unavoidable rounding at the corners. These result from the pressure differences created during the stretching.

With some special methods (carpet forming, flow molding, stamping), matched male and female mold halves are employed, with cooling conditions comparable to those found with injection molding.

Mold materials include the following:

- hard wood, for preproduction series;
- thermosets, for short runs, always with fillers;
- zinc alloys, for longer production runs;

- copper alloys, for longer production runs, electroformed mold;
- aluminum, for longer production runs, only forming molds;
- steel, for longer production runs, forming and stamping molds/dies.

For mass production, the molds are provided with means for temperature control. The mold temperatures vary between approximately 10°C (PP packaging), 70°C (PS, refrigerator liners), and up to 100°C (HDPE parts). Both single- and multicavity molds are found in machines, depending on the quantities to be produced.

a) Female Mold Methods

With female mold methods, the heated stock is pressed into the concave mold cavity (Fig. 7.1.3). While with straight vacuum forming there is only a pressure difference of 0.8 to 0.93 bar, compressed air machines usually work with forming pressures of 4 to 7 bar, and occasionally up to 25 bar. Since a higher force is available for forming, processing can take place at lower temperatures; this also saves on heating time. On the other hand, the energy requirements for compressed air are higher than for vacuum.

The wall thickness distribution is generally not improved by working with compressed air, but the corners and edges are better formed. Materials with a higher resistance to forming can be processed more easily. The closer contact between the cooling part and the mold improves the cooling action and thus shortens the cycle time.

In contrast to flat thermoformed parts, deeper ones are always formed with mechanical assistance because of the required uniformity in wall thickness. The use of plug assist assures that enough material is available for the bottom of the part so that the side walls and bottom are more uniform. Compared to simple negative forming, the wall thickness distribution is improved noticeably (Fig. 7.1.3).

The wall thickness distribution is determined by the stock temperature, plug shape and speed, plug temperature, plug stroke and timing of plug motion, mold motion, and start of forming air. With optimum adjustment of all parameters, a diameter:height ratio of 1:2 can be achieved with good material distribution. The plug material is an insulating substance such as hard felt or temperature-resistant plastic, which assumes a constant temperature

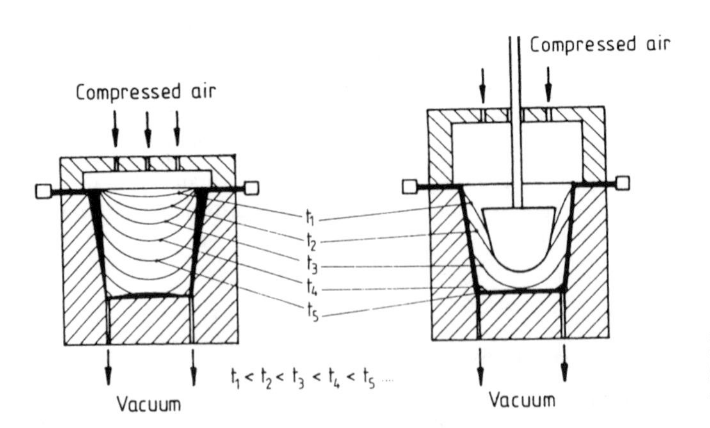

Figure 7.1.3 Air-assisted female mold with and without assisting plug.

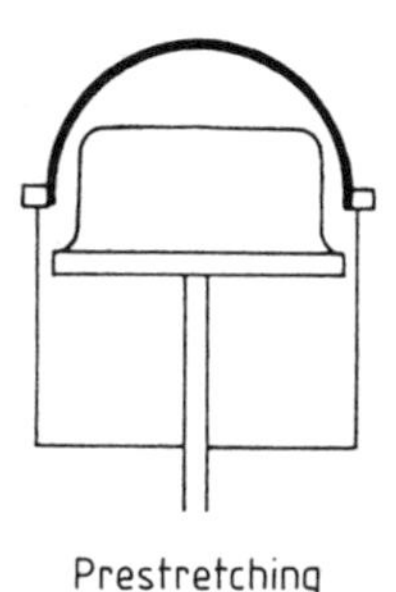

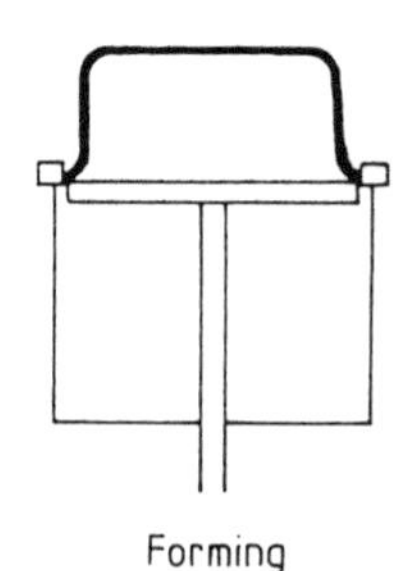

Figure 7.1.4 Male mold with prestretching.

after a few cycles, or heated metal plugs of aluminum or steel in order to avoid excessive cooling of the plastic stock.

b) Male Mold Methods

Male mold methods also find application often, e.g., for the production of instrument panels. With this method, the sheet is forced against a convex model (male mold) during the forming step, i.e., the plastic adheres completely and tightly to the male mold.

While a part with a thin bottom and thin corner results from simple female mold methods (without plug assist), the bottom remains thick and the upper rim region thinner with simple male mold methods. For this reason, the sheet is prestretched with male mold methods to extend it uniformly and thus avoid thick spots in the formed part (Fig. 7.1.4). With pronounced undercuts and complex molds, assisting plugs or slides are employed.

Cooling Process

By definition the cooling process begins upon contact of the sheet with the mold and ends with the start of the part removal process. In the majority of cases encountered in actual practice, the material is in contact with the mold on one side. On the other side, heat removal takes place via convection, either to the quiescent air (free convection) or with the aid of a blower (forced convection). To intensify the cooling, a water mist is sometimes added to the air flow with forced convection (a spray nozzle cooling unit).

Advantages and Disadvantages

Thermoforming, be it for production of packaging containers or large technical parts, is in direct competition with injection molding. Both methods can produce high-quality parts.

The obvious advantages of thermoforming over other production methods are the lower mold costs, the use of combination machines, rapid process sequence, and thus high output through the use of multicavity molds. In addition, the wall thicknesses achievable are lower than with injection molding. Most thermoformed parts generally weigh about 50% of the equivalent injection molded articles.

In contrast to injection molding, where the pressure can act on part formation only over the surface perpendicular to the flow direction, the entire part surface is utilized in thermoforming. This is the reason that the pressures are 2 to 3 orders of magnitude lower and large-area parts can be produced with lighter weight machines and lower forming pressures.

It must be mentioned, however, that the developments with regard to automation are already much more advanced with other molding methods. In addition, the part shape is subject to certain restrictions, i.e., as with injection molding, care must be taken to avoid undercuts that make part removal difficult.

7.2 Welding of Plastics

The plastics processor has available today a large number of *welding methods* with which a firm and durable bond can be achieved. Use of a particular method depends largely on the economics and achievable joint quality.

During the welding process, the material of the parts to be joined is heated to a flowable condition in the weld zone through the introduction of *energy (heat)* and both parts are joined together under *pressure*. This means that of the three material groups comprising thermosets, *thermoplastics*, and elastomers, only thermoplastics can be welded, since only they can be brought into a thermoplastic (melt) condition. Fig. 7.2.1 shows the temperature range in which the welding process takes place. It is identical to that in which the primary processing methods such as injection molding or extrusion occur.

The following physical processes are available to convert the material to be welded into a thermoplastic condition:

- conduction,
- convection,
- radiation,
- friction,
- induction.

Besides the introduction of energy into the weld zone, the pressure is another important condition. *Flow* in the *joint plane* as the result of pressure leads to entanglement of the molecular chains at the interface, thereby creating a *material bond* between the two parts.

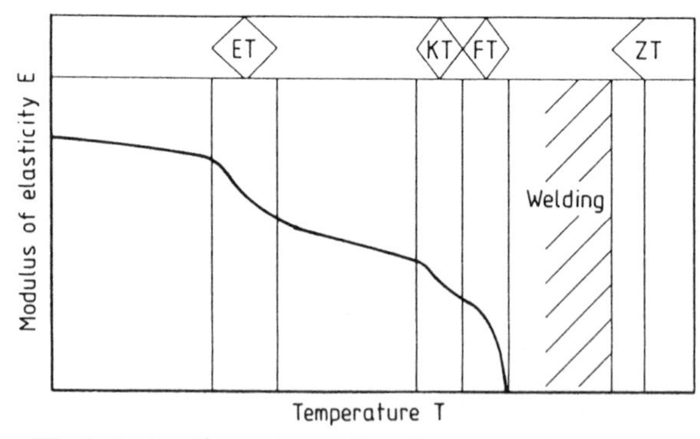

Figure 7.2.1 Temperature ranges for a semi-crystalline plastic (as example).

Conduction		Convection
Heated tool welding		Hot gas welding
Direct	Indirect	Hot gas welding
Heated tool butt welding Heated tool sleeve welding	Thermal contact welding Thermal impulse welding (one-sided and two-sided)	Fan welding High-speed welding (Draw welding) Hot gas extusion welding
Electric sleeve welding Heated wedge welding Bending with heated tool		**Radiation**
Friction		Heated tool radiation welding
Internal	External	
High-frequency welding	Rotation friction welding	Light beam welding Light beam extrusion welding Laser beam welding
Ultrasolic welding (direct and indirect)	Vibration welding	**Induction**
		Electromagnetic welding

Figure 7.2.2 Classification of welding methods on the basis of energy transfer means.

To achieve an adequate amount of melt in the weld zone, the duration of energy introduction (called the *heating time* or *welding time*, depending on the method) must be mentioned as a third important factor. The objective when optimizing parameters during welding is to determine the combination of pressure, time, and temperature that yields the highest quality result.

The welding methods are classified on the basis of the means used to introduce the energy. Following this principle, the common welding methods for thermoplastics are presented in Fig. 7.2.2.

All of the welding methods listed in Fig. 7.2.2 are based on the following steps for achieving a good weld joint:

1. Preparation of the joint surfaces: The joint surfaces must be dry, and dust- and grease-free. Any oxide layers (e.g., on polypropylene) must be removed by scrapping with a drag scraper.
2. Heating the joint surfaces: The heating time must be so selected that sufficient melt is available for the joining process.
3. Applying the joining pressure: Excessive joining pressure displaces too much melt from the joint zone so that colder material layers come into contact. Because of their lower

fluidity, there is no entanglement of molecular chains in the weld joint. This also occurs when there is only slight flow because the pressure is too low.

4. Cooling under pressure: To avoid warping and voids, the pressure should be maintained until the temperature has dropped so far that the joint achieves a rigid condition.
5. Finishing the weld seam (if necessary): If the weld bead that results with some methods is undesirable for practical (e.g., on window profiles) or visual reasons, it can be removed in a finishing step by means of cleaning or grinding.

Theoretically, all thermoplastics can be welded, but the structure and molecular weight of the material determine the technical feasibility to a large extent. Since welding takes place when the joint surfaces are in a plastic condition, this capability is an important criterion for the *weldability* of a plastic. Because of their structure and high molecular weight, materials such as polytetrafluoroethylene (PTFE), cast polymethyl methacrylate (PMMA), or ultrahigh-molecular-weight polyethylene (UHMWPE) cannot be plasticated at all, or only with difficulty, and thus are also technically difficult to weld.

Different thermoplastics can be welded together if the melting points are not far apart and the viscosity conditions in the melt are similar so that a material bond can be achieved under pressure. In addition to the surface tensions and the compatibility, the coefficients of thermal expansion and the shrinkage characteristics play a role. For this reason, only thermoplastics within the *amorphous* and *semi-crystalline* groups should be welded. The individual welding methods, the process technology, and the fields of application are presented in the following subsections.

7.2.1 Heated Tool Welding

All heated tool welding methods have in common that the energy is introduced to the joint surfaces by *conduction* with the aid of (usually electrically) heated metal tools. A distinction is drawn between direct and indirect heated tool welding. In *direct heated tool welding* the heat is introduced to the joint surface directly from the heated tool; in *indirect heated tool welding* heat flows to the joint surface through the part from the outside. Because of the poor thermal conductivity of plastics, indirect heated tool welding is employed only with very thin stock (films).

7.2.1.1 Direct Heated Tool Welding

Heated Tool Butt Welding

Heated tool butt welding is the most commonly employed thermal joining method for plastics. It is of great importance for joining *semi-finished products* (panels, pipes, window profiles) and is used increasingly in mass production for *molded part welding*. Its use for the joining of molded parts results from the possibility of being able to machine the heated metal tools to match even complicated, nonplanar joint surface contours. The heated tool surfaces are given a PTFE coating to prevent sticking.

Because of the service temperature limit of 240°C for PTFE, *heated tool radiation welding* is employed for the welding of high-temperature plastics and for plastics with very low melt viscosities. In this method, the part is held at a distance of approximately 1 mm from the heated tool and warmed by means of radiant heating (see Chapter 7.2.4).

To shorten the heating times for heated tool welding, *high-temperature heated tool welding* is employed. Heating takes place at temperatures of 300–500°C without a nonstick coating, and the adhering material is allowed to evaporate while the heated tool is not being used. Only those thermoplastics with a decomposition temperature in the specified range and which leave as little residue as possible upon evaporation find application with this method.

The process sequence consists of basically three phases, which are shown schematically in Fig. 7.2.3:

a) the *heating phase*,

b) the *change-over phase*,

c) the *joining phase*.

a) During the heating phase, the heat needed to melt the joint surfaces is introduced to the parts to be joined through direct contact with a heated tool at an appropriate temperature. There are two possibilities:

1. Heating with limited melting (*matching*): The parts to be joined are held against the heated tool with a high pressure (= the joining pressure) in order to eliminate any possible unevenness of the contact surfaces and assure complete contact. At this point some of the melt flows into the bead. Melting is interrupted after a certain period of time. This can be accomplished either with the aid of stops or by reducing the pressure to almost zero. *Pressureless warming* now begins.

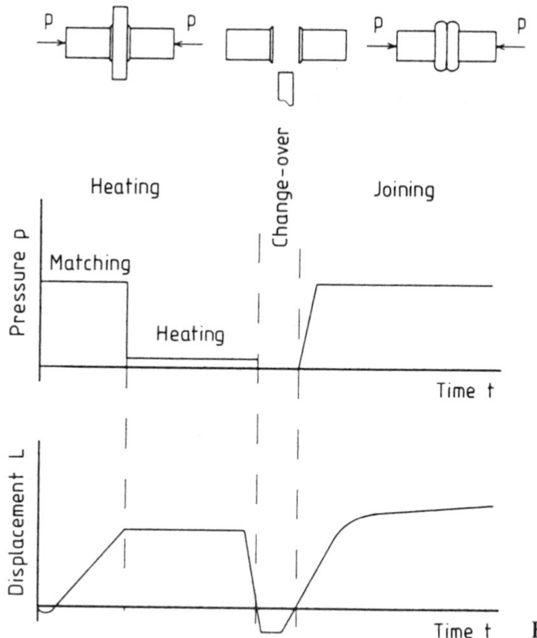

Figure 7.2.3 Heated tool butt welding.

2. Heating without limited melting: The melting process is not interrupted during the entire heating time so that melt flows continuously into the bead.

Generally, the limited melting technique is employed. Accordingly, the pressure and position (distance) as functions of time are presented only for this variation in Fig. 7.2.3. Heating without limited melting is employed only in special cases, e.g., with severely oriented parts or very thermally sensitive thermoplastics.

b) During the change-over phase, the heated tool is removed from the plane of the future joint. The temperature at the surface drops. The change-over time should be as short as possible to prevent excessive cooling of the surfaces and the associated reduced fluidity of the material.

c) During the *joining phase*, the actual bonding of the parts to be welded is accomplished by pressing the melted joint surfaces against one another. The flow caused by the joining pressure in the region of the weld results in entanglement of the molecules and thus the bonding of the two parts.

Heated tool butt welding is employed primarily for polyolefins. It represents the state of the art for joining PE and PP pipes for domestic water and gas lines. Because it is reproducible and capable of being automated, this method is also employed today for mass production. The spectrum of parts ranges from capacitors with a welded surface of 6×10 mm up to large shipping pallets.

Heated Wedge Welding

For lap welding of plastics sheets and webs (e.g., of PE, PP) or of coated fabric sheeting (e.g., of PVC) with a thickness greater than 0.3 mm, *heated wedge welding* is employed as a continuous joining technique (Fig. 7.2.4). While the heated metal wedge and *pressure roll* are guided by hand in manual welding, the heated wedge and transport/pressure rolls remain in one location in machine welding and the materials to be welded move past the heated wedge (sewing machine principle). When adjusting the welding parameters, care must be taken to achieve a combination of heated tool temperature and material speed that assures adequate and uniform melting of the joint surfaces.

Bending with Heated Tool

Bending with a heated tool (Fig. 7.2.4) is a welding technique for panels and sheets that is encountered more often in the manual trades. The wedge-tipped heated tool is pressed into the material up to three-quarters of its thickness and withdrawn after the heating time (almost pressureless). The panel is welded during the bending. To assure that adequate welding pressure can be applied, the tip angle of the heated tool should be approximately 20% less than the corner angle. To shorten the heating time with thick panels, a wedge-shaped groove can be milled beforehand.

Heated Tool Sleeve Welding

Heated tool sleeve welding is often employed as a joining technique for pipe runs. This creates a lap joint between the joint and sleeve (Fig. 7.2.5). One side of the heated tool has

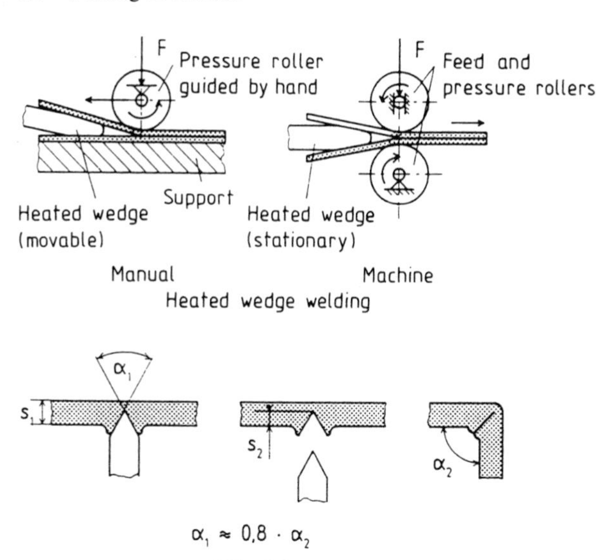

Figure 7.2.4 Heated wedge welding and bending with heated tool.

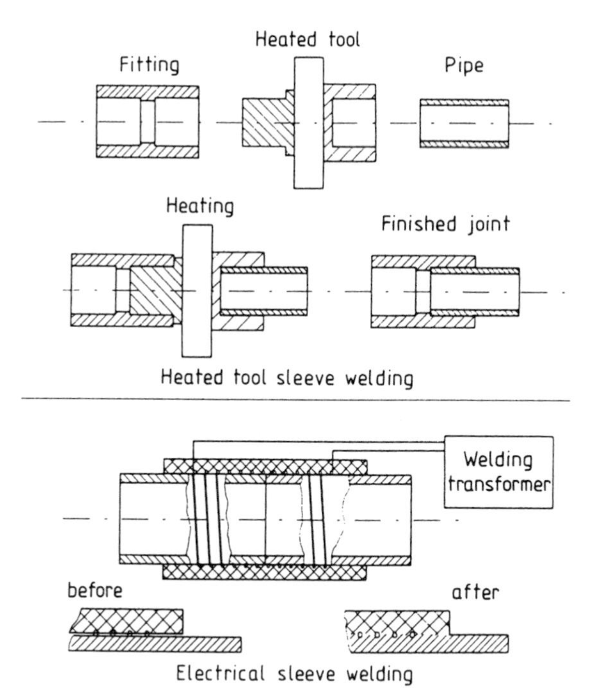

Figure 7.2.5 Heated tool sleeve and electrical sleeve welding.

the shape of the inside of the sleeve; the other side matches the outside shape of the pipe. The dimensions of the heated tool, pipe, and fitting are such that, after heating and change-over, a welding pressure is built up during the joining process.

This technique is employed almost exclusively for piping systems of PE and PP. Careful preparation of the surfaces to be welded is especially important when this technique is employed on the construction site.

Electrical Sleeve Welding

Electrical sleeve welding is an especially convenient method for the welding of piping. The inside surface of the injection molded sleeve contains electrical resistance wires, which are connected to a welding transformer after the end of the pipe has been positioned (Fig. 7.2.5). When a voltage is applied (the duration depends on the diameter) the resistance wires function as a heated tool. The inside surface of the sleeve and the outside surface of the pipe are melted. Since the sleeves are usually molded with frozen-in stresses, shrinking that supplies the necessary welding pressure occurs during the heating process. The resistance wires remain in the weld joint.

7.2.1.2 Indirect Heated Tool Welding Methods

Contact Welding

Contact welding is employed with thin parts ($\ll 1$ mm) and with films. A continuously heated tool with a nonstick coating conducts the heat to the joint surface through the part from the outside (Fig. 7.2.6). With this method the material is heated throughout.

Impulse Welding

Impulse welding is the most commonly employed indirect heated tool welding technique. It is used extensively with polyolefins in the packaging industry for sealing of bags and

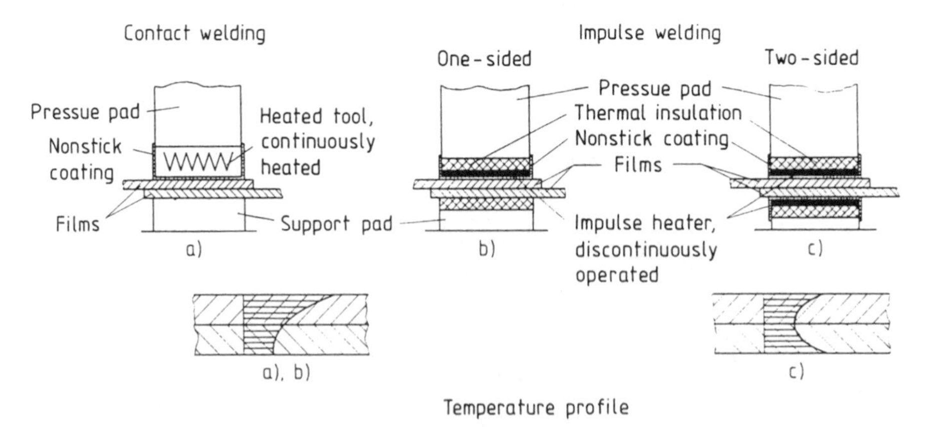

Figure 7.2.6 Contact and impulse welding.

sacks. As with contact welding, application of this technique is also limited to very thin parts (PE films: ≤3 mm; PVC films: ≤0.08 mm) because of the poor thermal conductivity of plastics. The advantage over contact welding is to be found in cooling under pressure.

At the start of the joining process, thin metal bars that have a nonstick coating (usually PTFE) are subjected to a brief but *high current pulse* and become heated. The heat is transferred through the film by conduction to the joint (Fig. 7.2.6). Two variations exist: *one-sided* and *two-sided impulse welding*. While the temperature profile in the parts to be joined is symmetrical because of the arrangement of the heating element in the two-sided variation, the profile is asymmetrical in the one-sided variation, just as with contact welding (Fig. 7.2.6). The process sequence with the indirect heated tool techniques results in a pronounced temperature profile, with the highest temperature always occurring on the surface. When establishing the welding parameters (heated tool temperature, contact time, magnitude and duration of current pulse), care must be taken to always satisfy the following condition: The temperature in the weld must be above the melting point and at the same time the surface temperature must not exceed the decomposition temperature.

7.2.2 Hot Gas Welding

Hot gas welding is usually performed by hand and thus requires a great degree of manual dexterity. The heat-conveying medium is generally hot, clean compressed air. With some oxidation-prone materials, inert gases such as nitrogen or carbon dioxide are employed. The surfaces to be joined are heated by the hot gases and welded together most often through use of a *filler rod* (Fig. 7.2.7). This technique is very well-suited for assembly and

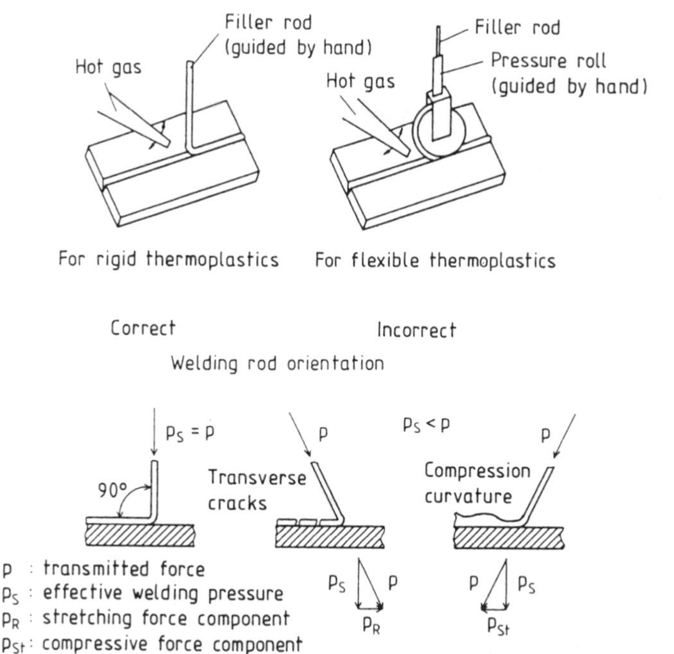

Figure 7.2.7 Hot gas welding with round nozzle.

repair work, and is used very often for *fabrication of containers and related apparatus*. Use of this technique is to be recommended whenever the weld joints are poorly accessible or materials are difficult to weld. Guidance of the gas nozzle and filler rod are, together with a constant pressure, reproducible *hot gas temperature* and constant *hot gas flow rate*, the most important factors affecting the *weld quality*. Three variations of the technique are in use: fan welding, high-speed welding (draw welding), and extrusion welding.

Fan Welding (Round Nozzle)

By moving the welding nozzle manually in a fan-shaped pattern, the parts to be joined and the welding material are heated. The welding rod is pressed into the prepared weld joint vertically by hand so that a "bow wave" forms ahead of the rod in the welding direction. With incorrect *filler rod positioning*, defects such as compression bulging or transverse cracks can result. While the filler rod is fed without any mechanical aids when welding rigid thermoplastics, use of a pressure roll is required when welding flexible materials (Fig. 7.2.7).

High-Speed Welding (Draw Welding)

With the *high-speed welding* (*draw welding*) technique, the welding material is fed through a guide into the hot gas nozzle (Fig. 7.2.8). By dividing the stream of hot gas, the *base* and *filler material* are heated and softened. The welding material, which issues from the guide in a softened condition, is pressed against the base material with the aid of the nozzle shoe. Higher welding speeds are achievable with this technique than with fan welding.

Extrusion Welding

Extrusion welding is recommended for joining thick-walled parts. While several layers of filler material are necessary with fan welding and high-speed welding, the welding gap is often filled in a single pass with extrusion welding. The base material is heated with hot gas, while the filler material is fed in a plasticized condition directly from the extruder and is pressed into the welding gap with the aid of a shoe (Fig. 7.2.8).

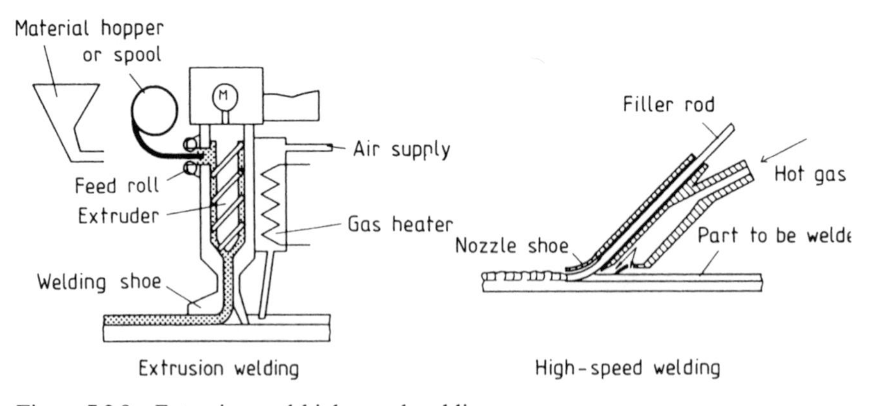

Figure 7.2.8 Extrusion and high-speed welding.

7.2.3 Friction Welding Methods

Friction welding methods utilize *frictional heat* to soften the surfaces to be joined. The methods are classified as utilizing either *external* or *internal friction.* While a *relative motion* of the parts to be joined with respect to one another generates the heat in the case of external friction, the *electrical* and *mechanical damping properties* of the materials as well as molecular vibrations generate heat "from the inside out" in the case of internal friction.

7.2.3.1 Methods Utilizing External Friction

Rotational Friction Welding (Spin Welding)

This method is suitable only for *rotationally symmetrical welds.* It is simple and inexpensive. The principle on which this method is based is illustrated in Fig. 7.2.9. The component held in the chuck is rotated, while the mating piece is kept stationary. The force F is applied through the rotating centers. As friction begins, the mating surfaces conform to one another and the temperature rises as a consequence of the frictional heat until the surfaces in contact begin to melt. Further heating is caused by shear processes in the melt. Once an adequate amount of melt has been generated as evidenced by the appearance of molten material at the outer edge of the joint, the stationary clamping mechanism is released, both components rotate together, and the weld cools under pressure. Reference values for the circumferential velocity and contact pressures are 0.8 to 2.5 m/s and 1 to 2 N/mm^2, respectively.

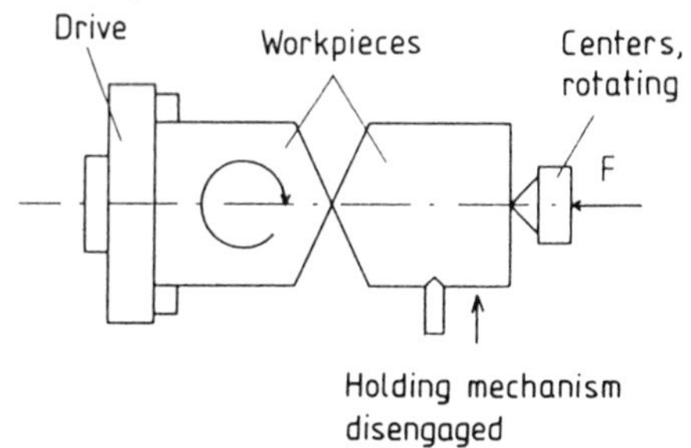

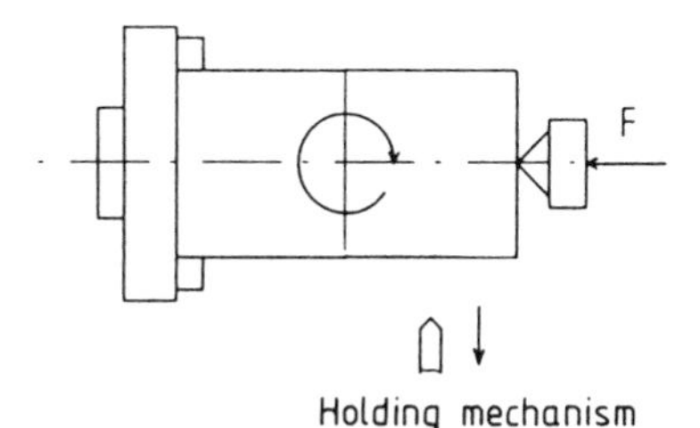

Figure 7.2.9 Spin welding.

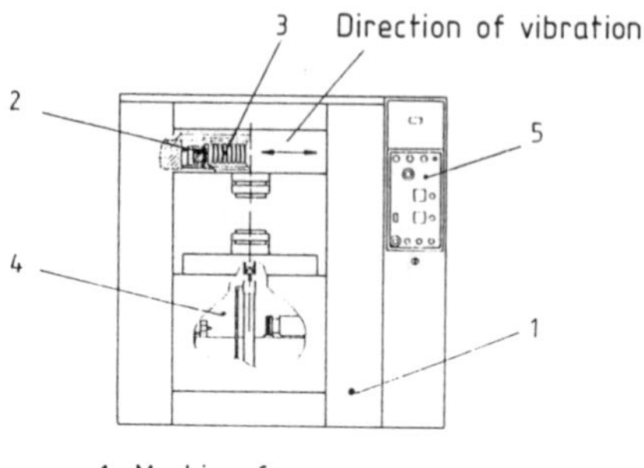

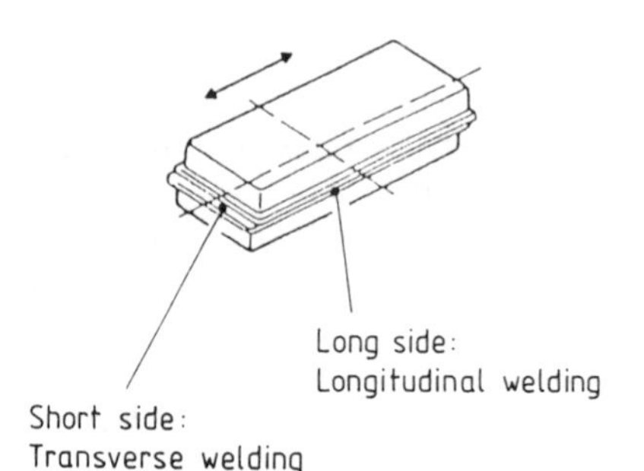

Figure 7.2.10 Linear vibration welding.

Vibration Welding

In contrast to rotational friction welding, *vibration welding* is suitable for irregularly shaped parts where the surfaces to be joined are not rotationally symmetrical. The *oscillating* frictional motion can be either *linear* or *angular* (Fig. 7.2.10). Welding takes place at frequencies between 70 and 300 Hz and pressures up to 4 N/mm^2. The amplitude of vibration can be up to 2.5 mm, with welding times of approximately 5 s.

Vibration welding has become well-established particularly among suppliers to the automotive industry. Significantly larger parts can be joined by means of vibration welding than with ultrasonic welding. Vibration welding has also proven useful for the welding of *hollow articles*, especially when the materials are not well-suited for indirect ultrasonic welding because of their high damping.

7.2.3.2 Methods Based on Internal Friction

Ultrasonic Welding

In *ultrasonic welding* the *mechanical damping properties* of plastics are utilized. Mechanical vibrations in the ultrasonic range (20–40 kHz; in special cases up to 70 kHz) are transformed into heat by internal friction and interfacial friction at the surfaces to be joined. The softened material is then joined under pressure (Fig. 7.2.11).

A high-frequency generator converts the line frequency into high-frequency electrical vibrations. These are then usually converted by a piezoelectric transducer into mechanical *longitudinal vibrations* that are transmitted to the component to be welded via a welding tool (sonotrode) and booster. Amplitudes of vibration range from 20 to 100 μm. Reflection

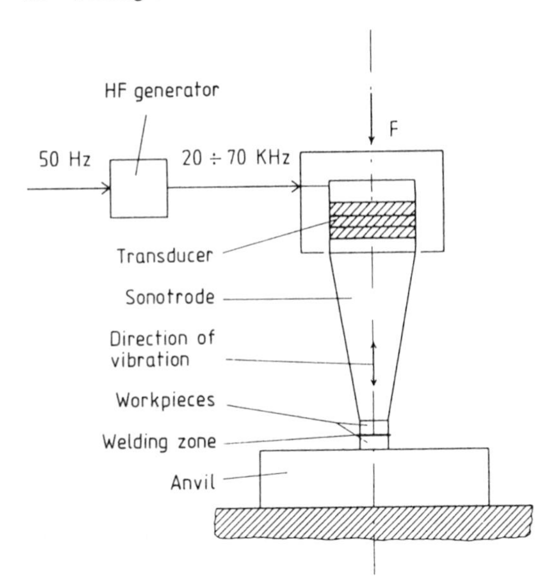

Figure 7.2.11 Ultrasonic welding.

from the anvil creates a *standing wave*. Since the vibrating assembly (transducer, booster, sonotrode) is operated in *resonance*, special attention must be given to the geometric design. There are two variations of this technique: *direct* and *indirect ultrasonic welding*. The distance between the face of the sonotrode and the joint surface serves as the differentiating criterion. If the distance is less than 6 mm, the method is called *direct ultrasonic welding*, while for distances greater than 6 mm the method is considered *indirect ultrasonic welding*. Generally, all thermoplastics can be welded with the direct method. Only rigid thermoplastics are suitable for the indirect method, since in flexible thermoplastics the ultrasonic energy is absorbed to such an extent because of the high damping as it passes through the material that hardly any warming of the joint surface is achieved.

Because of the very short welding times (0.2–1.5 s), the method is very well-suited for mass-produced articles, e.g., for long production runs at suppliers to the automotive industry, for household goods, electrical appliances, and toys. The method is used extensively in the packaging industry for welding (sealing) of films. Additional fields of application include ultrasonic inserting, riveting, and staking as well as spot welding.

High-Frequency Dielectric Welding

Some plastics contain *polar groups* that can become aligned in an external electric field. As a result, these polar plastics (such as PVC, ABS, and PA, for instance) can be made to vibrate on the molecular level when placed in an alternating electric field. The intermolecular friction then heats the material. This heating can be maintained until the plastic is transformed into the melt state. The electrical polarity of a material is indicated by its *dissipation factor* tan δ. The higher the value of tan δ, the greater is the polarity and the

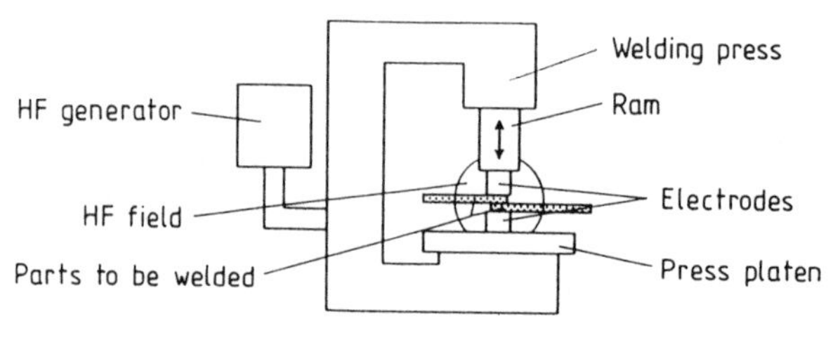

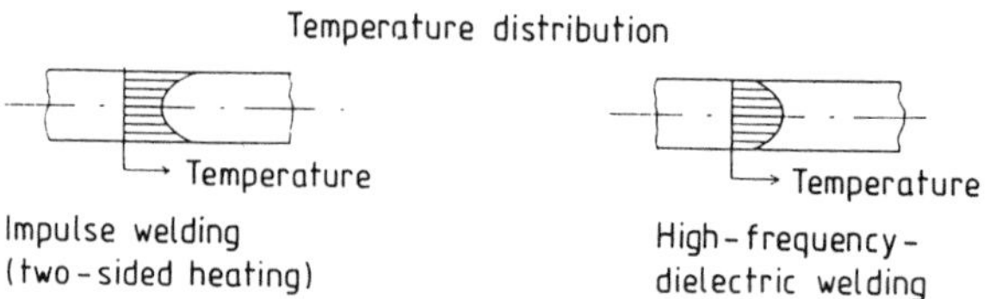

Figure 7.2.12 High-frequency dielectric welding.

better the plastic can be heated in an HF field. The following table contains tan δ values for several plastics:

Material	tan δ
Rigid PVC	0.03–0.02
Flexible PVC	0.1–0.05
ABS	0.03–0.01
PA	0.04–0.02
PS	0.0008–0.0003
PE	0.0003–0.0005

The principle employed for HF welding is illustrated in Fig. 7.2.12. An HF generator converts the line current into a high-frequency alternating current. The upper capacitor plate is the so-called electrode and conforms to the desired weld joint. The lower capacitor plate functions as the lower platen.

An important aspect of HF welding is that the heat is generated in the material itself. The electrodes normally remain cold and in comparison to impulse welding, for instance, a very beneficial temperature distribution results (Fig. 7.2.12).

This method is employed primarily for joining flexible PVC film or PVC-coated fabric, and is used extensively by the automotive industry and bag makers as well as the toy and sporting goods industries. Often the tools have cutting edges or profiles that trim the material next to the seam and/or simultaneously emboss the seam.

7.2.4 Radiation-Based Welding Methods

The welding methods based on the introduction of energy by means of radiation can be divided into two groups: methods that utilize the thermal radiation of a heated tool and methods that operate with the aid of *light beams* or *laser beams*.

Heated Tool Radiation Welding

With heated tool welding, stringing may occur during the change-over phase with some materials in spite of the nonstick coating. A process in which the components are no longer in direct contact, but rather are at a slight distance (0.5–1 mm) from the heated tool, provides a remedy. Now the heating takes place primarily via radiant heat, with a certain amount of convection as well. Since the components expand as they are heated, the distance from the heated tool must be continually corrected to prevent contact.

Light Beam Welding

Noncontact melting and welding of plastics is also possible through the use of focused light beams. This method is practical only for *opaque* materials, since in transparent material the absorption of light is insufficient as a rule. A distinction is made between *direct* and *indirect light beam welding.*

In the *direct* method the radiant energy impinges directly on the surfaces to be joined, while in the *indirect* method the energy must be transmitted through the components to the surface to be joined. To date, light beam welding has been employed only in special situations involving film.

Light beam extrusion welding is another variation. Here the filler material is supplied by an extruder, as in conventional extrusion welding, but preheating of the components is no longer accomplished by means of hot gas, but rather by means of focused light beams. The material is distributed and pressed into the weld joint once again with the aid of shoes.

Laser Beam Welding

Laser beam welding is becoming increasingly common, especially for film. The high energy density of the laser beam makes high welding speeds possible. At the same time, however, there is the problem of creating an adequate volume of molten material because of the low thermal conductivity of plastics together with the high risk of material degradation at the surface.

7.2.5 Induction Welding (Electromagnetic Welding)

The principle of induction heating is employed with the *induction* (or *electromagnetic*) welding technique. A filler material containing a magnetic powder is placed between the components to be joined. The filler material is usually the same or similar to that of the components. The necessary thermal energy is provided by a high-frequency magnetic field that heats the particles through hysteresis and eddy current losses, thus raising the filler material to the melting point. This heat is then transmitted to the surfaces to be joined, which are held together under pressure (Fig. 7.2.13).

This technique is similar to high-frequency welding in that the heat is generated in the material itself and surfaces that are sometimes poorly accessible can be welded. A disadvantage is the poor conversion of energy in the filler material so that generator frequencies of up to 10 MHz must be employed.

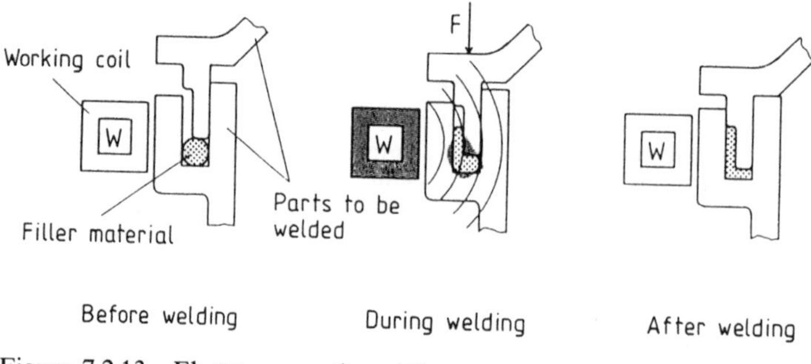

Figure 7.2.13 Electromagnetic welding.

7.3 Adhesive Bonding of Plastics

Adhesive bonding of plastics has gained enormously in importance as a joining technique for entire surfaces for the purpose of transmitting force and represents a welcome addition to welding. While only thermoplastics can be welded, adhesive bonding is a joining technique applicable to *all plastics*, i.e., *thermoplastics* as well as *elastomers* and *thermosets*. Adhesive bonding can be employed to join not only similar, but also dissimilar, materials.

The following advantages favor the use of adhesive bonding:

- a *uniform force distribution* over the bonded surface permits utilization of the material thickness up to the load limit (weight savings);
- thin, small, and even complexly shaped parts can be bonded;
- the adhesive bond can also fulfill additional functions such as *sealing*, leveling of unevenness through use of a filling adhesive, *vibration damping*, and thermal or electrical *insulation*;
- adhesive bonds give the designer wide latitude with regard to design, materials selection, and fabrication technique for the components to be joined.

The following limiting factors, however, must also be mentioned with regard to use of this technique and fields of application:

- *setting* and *curing* require a certain period of time in the production sequence;
- sometimes a complex *pretreatment* of the surfaces to be bonded is necessary;
- under certain circumstances there is poor *aging behavior* in the presence of chemical substances.

7.3.1 Mechanism of Adhesive Bonding

The *strength* of an adhesive bond results from *adhesion* and *cohesion*. *Cohesive forces* are responsible for the bonding within a material, i.e., the adhesive as well as the component to be bonded (*substrate*). *Adhesive forces* are responsible for the bonding between the adhesive and the component (Fig. 7.3.1).

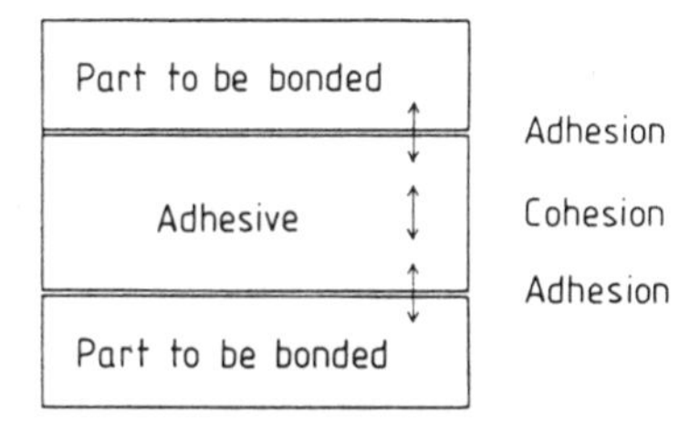

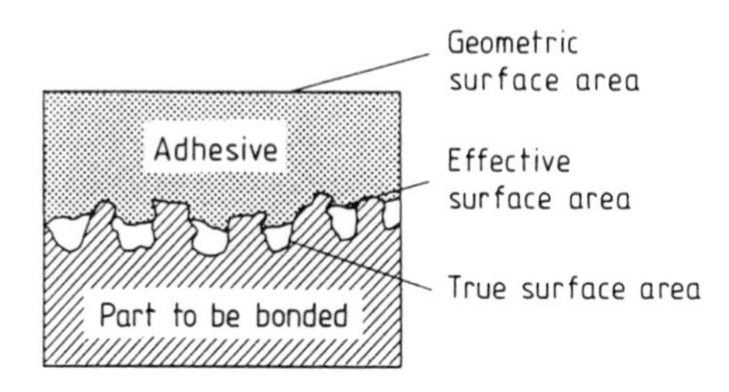

Figure 7.3.1 Designation of forces and surface areas for adhesive bonding.

A distinction is made between mechanical and specific adhesion. *Mechanical adhesion* results from the mechanical anchoring of the adhesive in pores of the component to be bonded. It is important for materials such as foams and cork. Adhesive bonds resulting from *intermolecular secondary valence forces* are attributed to specific adhesion.

Both cohesive and adhesive forces are intermolecular forces with a very limited range (a few nanometers). Adhesive thus must be within immediate proximity of the component surface for the adhesive forces to be effective. This is the reason for the stringent cleanliness requirements when employing adhesive bonding. Dirt particles or grease on the surface reduce the bond strength substantially.

The *surface structure* of the surfaces to be bonded has a significant influence on the bond strength that can be achieved in practice. A distinction is made between the *true surface area*, which is determined by the surface roughness, and the *geometric surface area*, which is determined by the component dimensions. The interface between the adhesive and the component is described by the *effective surface area*, which results from the degree of wetting of the true surface area (Fig. 7.3.1). The adhesive is usually applied in liquid or plastic form in an attempt to achieve complete wetting and thus assure close contact between the adhesive and the component.

7.3.2 Classification of Adhesives

A practical classification of adhesives is based on the type of setting mechanism. This yields two large groups (Fig. 7.3.2):

- physically setting adhesives,
- chemically setting (reactive) adhesives.

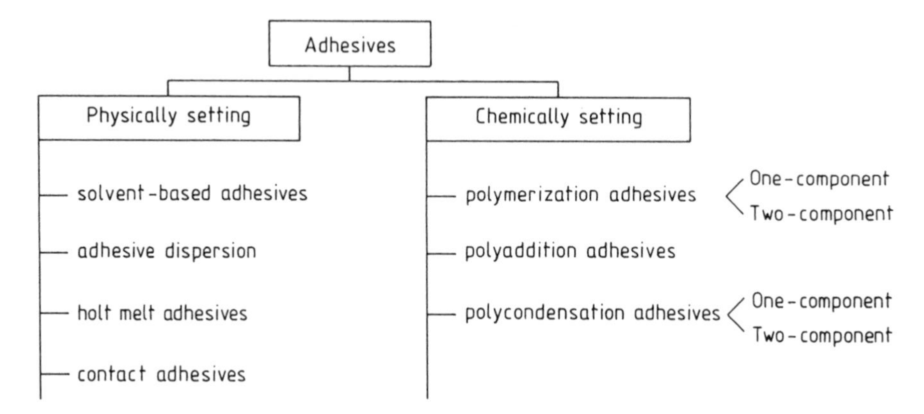

Figure 7.3.2 Classification of adhesives.

7.3.2.1 Physically Setting Adhesives

Solvent-Based Adhesives and Adhesive Dispersions

To achieve good wetting, adhesives are often dissolved in organic solvents or dispersed in water. Prior to bonding, the solvent or dispersing agent must be removed. This is accomplished via evaporation or absorption by the substrate. When employing solvents, care must be taken to assure compatibility with the substrate as otherwise detrimental changes to the components can result (e.g., stress cracking, etc.). Bonding with a pure solvent (solvent bonding) represents a special variant of a solvent-based adhesive. This type of joining can be employed only with thermoplastics that dissolve readily and is called *diffusion bonding*.

Hot Melt Adhesives

Hot melt adhesives are applied as melts or a softened mass. Bonding takes place immediately after application, before the temperature drops below the melting point or flow temperature. It is also possible to allow the adhesive to solidify and then remelt it immediately before bonding. Since cooling from the thermoplastic condition takes place relatively rapidly, the joint must be held together for only a short period of time, which makes this technique economically interesting for mass production in particular. There are also chemically setting adhesives that must be melted when used, but these are classified as reactive adhesives.

Contact Adhesives

For technical applications, contact adhesives usually contain solvents that must be evaporated prior to bonding (*open time*). Bonding can take place only after the adhesive is dry to the touch. Once the components have been brought into contact, no further corrections are possible.

Contact adhesives without solvents (*pressure-sensitive adhesives*) find application on labels, adhesive bandages, adhesive tape, etc. Because of the low cohesive forces, most bonds involving pressure-sensitive adhesives can be separated numerous times.

7.3.2.2 Chemically Setting Adhesives (Reactive Adhesives)

The bonding principle of reactive adhesives is based on the formation of macromolecules after joining as the result of a *chemical reaction.* These are the same reactions used to synthesize plastics, namely *polymerization, polyaddition*, or *polycondensation.* Usually *cross-linked macromolecules* are formed. The reaction can be initiated by so-called reagents (*hardeners, accelerators*), by *heat*, or even by *moisture in the air* (e.g., cyanoacrylate-based adhesives).

If reagents are employed, the adhesive is called a two- or multicomponent system (mixed adhesives). With these systems, the adhesive must be mixed in specified proportions prior to application. Careful observance of these proportions is a prerequisite for complete hardening. After mixing, the adhesive can be used for only a limited time, the so-called *pot life.* The pot life ends when the mixture exhibits an initial temperature rise caused by the chemical reaction. It depends on the composition of the mixture (Fig. 7.3.3) and is affected also by the ambient temperature. The lower the temperature, the slower chemical reactions take place.

7.3.3 Material Effects on the Bonding Capability of Plastics

The wide variety of polymeric materials means from the adhesive bonding standpoint that not all plastics can be adhesively bonded equally well with the same adhesives or with the same bonding technique. There are basically two reasons for this:

1. The chemical structure of the plastics (components to be bonded and adhesive): Depending on the chemical structure, a distinction is made between *polar* and *nonpolar* plastics. The polarity is responsible for the existence of *electromagnetic interactions.* The chemical structure also determines the *surface tension* of the materials, which in turn is a measure of the *wetting properties.* An additional important variable is the solubility of the components to be bonded, which can result in diffusion of the adhesive into the component and thus the formation of additional bonding forces between the component and the adhesive. The effect of these chemical structure-based properties on the bonding

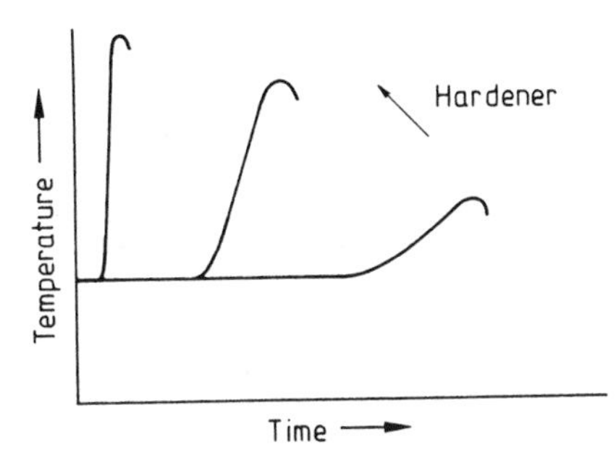

Figure 7.3.3 Reaction curves for two-component adhesives.

Plastic	Wettability	Polarity	Solubility	Adhesive properties
Polyvinyl chloride Rigid	+	Polar	Soluble	+
Polystyrene	--	Nonpolar	Soluble	+
Polyethylene	--	Nonpolar	Insoluble	--
Polyamide 66	+	Polar	Poorly soluble	–
Polymethyl methacrylate	+	Polar	Soluble	+
Phenol formaldehyde molding compound Typ 31	+	Polar	Insoluble	+
Unsaturated polyester molding compound Typ 801	+	Polar	Insoluble	+

Good : +
Moderate : –
Poor : --

Figure 7.3.4 Properties important for adhesive bonding.

characteristic is shown in Fig. 7.3.4 for a select number of materials. Polarity, good wetting properties, and solubility promote good bonding.

2. The mechanical deformation behavior of materials: Plastics generally exhibit significant deformation when subjected to mechanical stress. These deformations must be absorbed by the adhesive bond. Accordingly, the mechanical deformation behavior of the adhesive and the component to be bonded must be matched to one another to assure that, if possible, there are no stress irregularities that could lead to premature failure of the adhesive bond.

7.3.4 Process Sequence for Adhesive Bonding

The processing technique is of decisive importance on the production of high-quality adhesive bonds. Regardless of the type of adhesive employed, the bonding process can be divided into the following steps:

- *Designing* suitable joint surfaces: The most important prerequisite for adhesive bonding is that the components to be bonded and the joints be properly designed for bonding. The most desirable type of load is a shear stress; tensile or peeling loads should be avoided. If the risk of peeling exists, reinforcements to reduce the load, for instance, must be provided. Figs. 7.3.5 and 7.3.6 illustrate some of the common adhesive joints.
- *Cleaning* the surfaces to be bonded: The surfaces to be bonded must be thoroughly cleaned to remove foreign matter such as grease, dust, mold releases, water, etc. This is accomplished in continuous baths with organic solvents or alkaline cleaning agents, ultrasonic

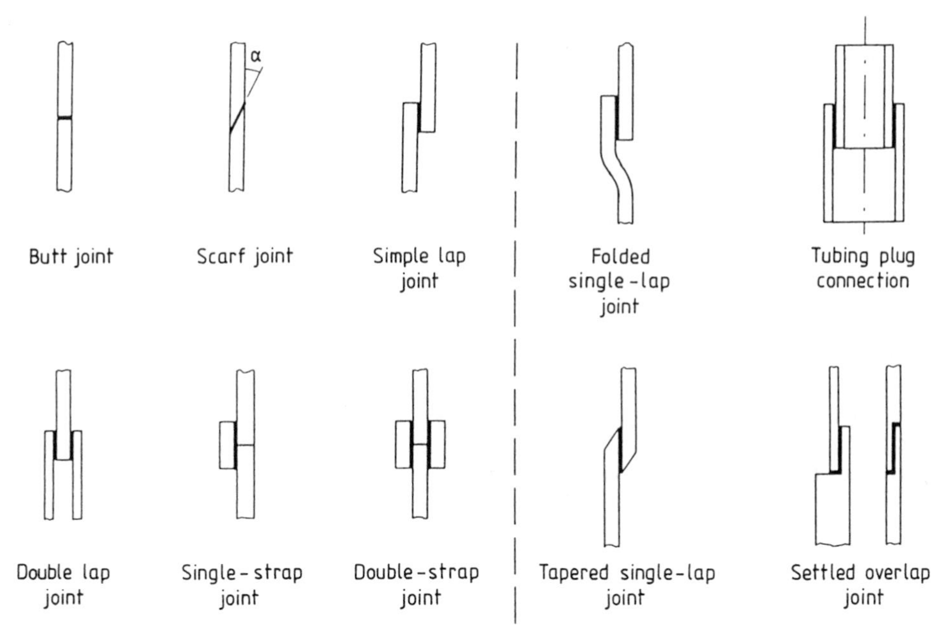

Figure 7.3.5 Types of adhesives joints 1.

Figure 7.3.6 Types of adhesives 2.

or steam degreasing baths. For a number of plastics, bonding can take place immediately after cleaning. For other plastics, the following step is necessary:

- *Treating* the surfaces to be bonded: There are two possibilities for treating the surfaces to be bonded: roughening and changing the surface structures and properties. Roughening can be accomplished mechanically (grinding, sandblasting) or chemically (etching). Roughening is sometimes important with compression molded parts of phenol-formaldehyde or melamine-formaldehyde resins as well as for components of glass-fiber- reinforced unsaturated polyester or epoxy resins, since the skin contains adhesive-repelling components that cannot be removed with cleaning agents. Of the mechanical roughening methods, sandblasting is preferred because of its uniformity. With hard-to-bond plastics—such as polyolefins or polyacetals—it is necessary to change the surface structures and properties in order to activate the surfaces. This usually involves an oxidizing step (flame or corona discharge treatment) or partial oxidation (etching) to render the surface receptive to the adhesive.
- *Applying* the adhesive: When applying the adhesive, care must be taken to assure a constant thickness of adhesive film and uniform wetting of the joint surfaces. For one-off or short-run production, the adhesive is applied with a brush or spatula. For mass production, a machine applicator is employed. When using multicomponent systems, homogeneous mixing of the components must be assured prior to application.
- Waiting until the adhesive is ready for bonding: Depending on the type of adhesive, certain waiting periods must be observed before the components are brought together. For instance, when using solution adhesives, care must be taken to assure that the evaporation time (time required for the solvent to evaporate) is observed. When bonding

with solvent cements, the waiting time (open time) is limited in order to assure that the solvent cement is still effective.

- *Joining and holding*: Often, pressing the components by hand or with a roller is sufficient. With larger parts, presses are employed. The pressure is intended to displace the air between the surfaces to be joined. With so-called *quick-acting adhesives*, maximum bonding properties are obtained by briefly applying a high pressure. As a rule, physically or chemically setting adhesives require fixtures to hold the parts together while the adhesive sets. An increased temperature generally accelerates the setting process. The contact pressure also affects the adhesive film thickness.
- *Hardening/curing of the adhesive.*
- *Removing the holding fixture*: After parts have been adhesively bonded, it often is necessary to wait a certain period of time before a load can be placed on the adhesive joint. This waiting period can be as long as several days. When properly executed, adhesive bonding of plastics must be considered a proven joining technique today for the production of permanent bonds capable of withstanding high forces.

7.4 Machining

The machining of plastics parts is often unavoidable in spite of the modern production methods for molded parts, such as injection or transfer molding, for instance. Molded parts must be finished (deflashed); complicated individual parts must be drilled, ground, or finish-turned, for instance; and semi-finished goods must be cut to size. For all of the above, the properties of plastics as a material must be taken into consideration. Since the *thermal conductivity* of plastics is very low, the *cutting speed, feed rate*, and *tool geometry* should be so selected that the resulting heat is removed with the chip for the most part. During machining, thermoplastics and thermosets must not exceed 60°C and 150°C, respectively. If cooling is necessary, care must be taken to assure that the *coolants* used do not lead to dissolution or swelling of the material. Machining is performed with tools of *high-speed steel, super-speed steel* (HSS), *metal carbide/hard metal* (sintered metal carbide alloy), *diamond*, and *corund.*

7.4.1 Sawing

Circular saws are suitable for *straight cuts.* Only high-speed steel or carbide-tipped blades can be employed, and they must be hollow-ground and have no set (Fig. 7.4.1). For thin sheets/panels and hard materials, a small *pitch* is required. The circular saw motors should be sized larger than for woodworking. Table 7.1 contains some guidelines for *sawing* of plastics.

With *brittle plastics* in particular, the saw table should be so adjusted that the teeth protrude just past the surface of the sheet to be sawn to avoid *chipping* of the material as the blade passes through it, thereby assuring a clean cut.

Pipes, blocks, and thick panels are cut to size with a *band saw*. The band saw is also recommended for *curved cuts.* The teeth of band saw blades generally have a slight set to prevent *clogging* by the swarf. The interrelationship between optimum *tooth pitch, cutting*

Table 7.1 Values from VDI Guidelines 2003 for the Sawing of Plastics

	Plastic		Abbreviation	Tool	α	γ_K	γ_B	t	v_K	v_B
Thermosets	Laminates and compression molding compounds	Organic fillers	Hp Hgw GFK	SS HM	30 to 40 10 to 15	5 to 8 3 to 6	5 to 8	4 to 8 8 to 18	to 3000 to 5000	to 2000
		Inorganic fillers		HM Diamond	Diamond grit Z				1000 to 2000	300
Thermoplastics	Polymethyl methacrylate and copolymers Polystyrene and styrene-acrylonitrile copolymers Acrylonitrile/butadiene/styrene copolymers Styrene butadiene copolymers Polyoxymethylene Polyacetal Polycarbonate Polyfluoroethylenes Polyvinyl chloride (rigid) and VC copolymers Cellulose ester Polyolefins Polyamides (nylons)		PMMA AMMA PS SAN ABS SB POM PC PTFE PVC CA CAB PE, PP PA	SS	30 to 40 SS 10 to 15 HM	0 to (−)4 0 to 5 5 to 8 SS 0 to 5 HM	0 to 8	2 to 8* *use closely spaced teeth for brittle materials		to 3000

v = cutting speed in m/min
K = index for circular saws
B = index for band saws

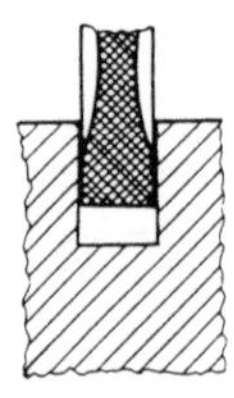

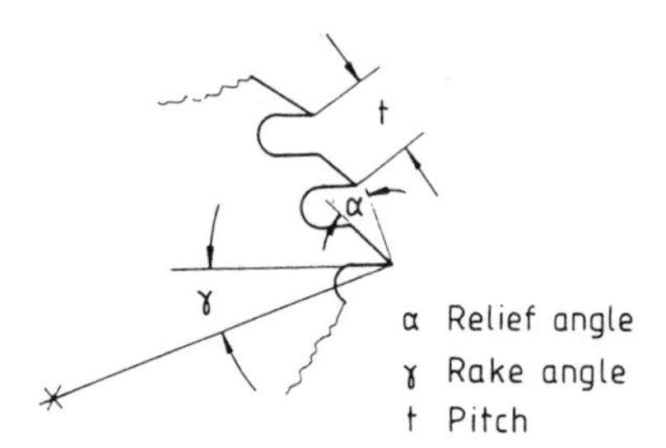

Figure 7.4.1 Tooth shape and tool angles for saw blades.

speed, and *stock thickness* depends on the type of material. Table 7.1 contains more detailed information in this regard.

7.4.2 Milling

In contrast to milling cutters for metals, *milling cutters for plastics* have a lower number of cutting edges, while compared to cutters for woodworking, however, the number of teeth is greater, except for the single-edged top millers. Millers are made of high-speed steel or

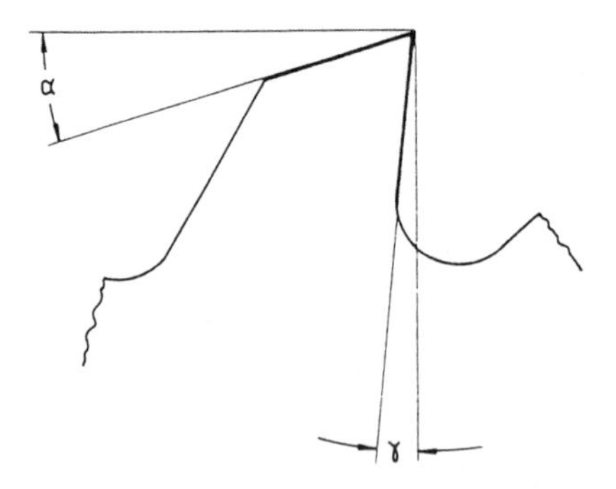

Figure 7.4.2 Tool angles for milling cutters.

sintered metal carbide, but can also be carbide-tipped. The *tool angles* for milling cutters are shown in Fig. 7.4.2. The *cutting speed* should be as high as possible and the *feed rate* relatively low in order to prevent chipping of brittle materials. Appropriate *cooling* is important. For *edge milling* and *shaping* of thermoplastics, the tool should not have more than two cutting edges to keep the chip space (gullet) sufficiently large and *vibrations* resulting from the number of teeth low. The highest *milling rates* are achieved with *single-edged tools* when high *surface quality* is desired. *Face milling* is more economical than edge milling for *flat surfaces.*

In general, the following rule applies: The harder the material and the thinner the workpiece, the closer the *radial rake angle* approaches zero from positive values. The softer the material, the lower is the number of cutting edges, while the feed rate is higher in order to remove heat with the chip. Table 7.2 contains more detailed information.

Table 7.2 Values from VDI Guideline 2003 for the Milling of Plastics

	Plastic		Abbreviation	Tool	α	γ	v
Thermosets	Laminates and compression molding compounds	Organic fillers	Hp Hgw GFK	SS HM	to 15 to 10	15 to 25 5 to 15	to 80 to 1000
		Inorganic fillers		HM	to 10	5 to 15	to 1000
				Diamond	Diamond concentration approximately 50%; grain size D125 to 250		to 1500
Thermoplastics	Polymethyl methacrylate and copolymers		PMMA AMMA	SS	2 to 10	1 to 5	to 2000
	Polystyrene and styrene-acrylonitrile copolymers		PS SAN				
	Acrylonitrile/butadiene/styrene copolymers		ABS				
	Styrene butadiene copolymers		SB				
	Polyoxymethylene Polyacetal		POM		5 to 10	to 10	to 400
	Polycarbonate		PC		5 to 10	to 10	to 1000
	Polyfluoroethylenes		PTFE				
	Polyvinyl chloride (rigid) and VC copolymers		PVC		5 to 10	to 15	to 1000
	Cellulose ester		CA CAB		5 to 25		
	Polyolefins		PE, PP		5 to 15	to 15	to 1000
	Polyamides (nylons)		PA				

v = cutting speed in m/min
The feed rate can be up to 0.5 mm/revolution

7.4.3 Grinding and Polishing

Grinding rarely is employed in plastics processing. *Finishing* or *smoothing* of saw cuts can be accomplished with the aid of commercially available *abrasive papers* (from coarse to extremely fine grains) or with *belt sanders*. The *speed* of belt sanders should be about 10 m/s. *Thermosets* generally can be processed well by *grinding/sanding*, while *thermoplastics* tend to "smear" severely because of the high amount of heat generated. Silky to mirror-finish surfaces can be achieved by means of *polishing*. The polishing can be performed by hand or machine with *felt* or *buffing wheels*. The polishing agents employed have a decisive influence on the *surface quality*. A prerequisite for good polishing is that the surface *irregularities* to be removed (scratches, grooves) be only minimal. When polishing thermoplastics, care must be taken to assure that the surfaces are not heated excessively or melt as a result of the *frictional heat* generated. This can be achieved by interrupting the polishing process several times.

7.4.4 Drilling

Drilling is encountered very often in the shop and in production. The *hand-held drills* or *drill presses* that are found almost everywhere can be employed for this purpose. The *twist drills* used for metals can almost always be used for plastics. Twist drills with a *high helix* are more practical, since they assure good *chip removal*. The *point angles* are generally 60° to 90° for thermoplastics. The tool angles for twist drills are shown in Fig. 7.4.3. More detailed information on the cutting edge geometry is listed in Table 7.3 for various materials.

Because of the severe *expansion* of plastics as a result of the *frictional heat* generated during drilling, final hole dimensions are 0.05 to 0.1 mm smaller than the drill bit diameter. Accordingly, an appropriately *oversized drill bit* is employed in actual practice.

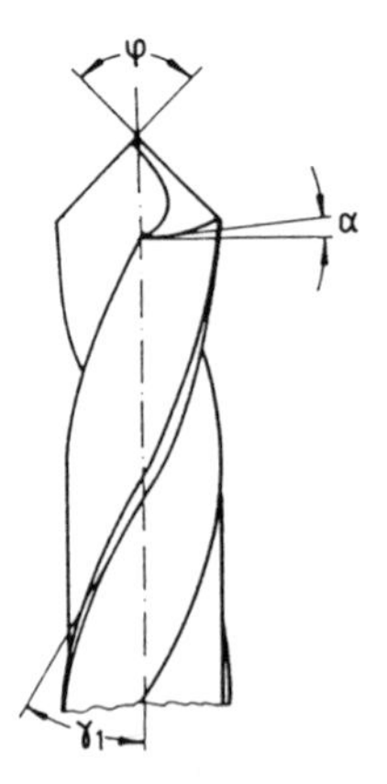

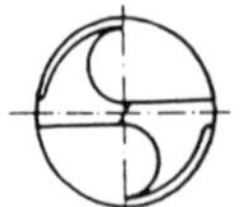

α Relief angle
γ_1 Helix angle
φ Point angle

Figure 7.4.3 Tool angles for twist drills.

Table 7.3 Values from VDI Guideline 2003 for the Drilling of Plastics

	Plastic		Abbreviation	Tool	α	γ	φ	η	s
Thermosets	Laminates and compression molding compounds	Organic fillers	Hp Hgw GFK	SS HM	6 to 8 6 to 8	6 to 10 6 to 10	100 to 120 100 to 120	30 to 40 100 to 120	0.04 to 0.6 on dr diameter and filler
		Inorganic fillers		HM Diamond	6 to 8	0 to 6	80 to 100	20 to 40	
Thermoplastics	Polymethyl methacrylate and copolymers		PMMA AMMA	SS	3 to 8	0 to 4	60 to 90	20 to 60	0.1 to 0.5
	Polystyrene and styrene-acrylonitrile copolymers		PS SAN	SS	3 to 8	3 to 5	60 to 90	20 to 60	0.1 to 0.5
	Acrylonitrile/butadiene/styrene copolymers		ABS	SS	5 to 8	3 to 5	60 to 90	30 to 80	0.1 to 0.5
	Styrene butadiene copolymers		SB	SS	8 to 10	3 to 5	60 to 75	30 to 80	0.1 to 0.5
	Polyoxymethylene Polyacetal		POM	SS	5 to 8	3 to 5	60 to 90	50 to 100	0.1 to 0.5
	Polycarbonate		PC	SS	5 to 8	3 to 5	60 to 90	50 to 120	0.2 to 0.5
	Polyfluoroethylenes		PTFE	SS	16	3 to 5	130	100 to 300	0.1 to 0.3
	Polyvinyl chloride (rigid) and VC copolymers		PVC CA	SS	8 to 10	3 to 5	80 to 110	30 to 80	0.1 to 0.5
	Cellulose ester		CAB	SS					
	Polyolefins Polyamides (nylons)		PE, PP PA	SS	10 to 12	3 to 5	60 to 90	50 to 100	0.2 to 0.5

v = cutting speed in m/min
s = feed rate mm/revolution
The helix angle b of the drill should be approximately 12–16°

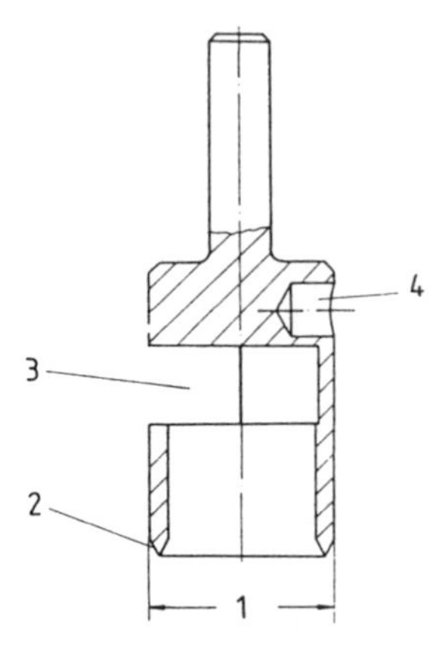

Figure 7.4.4 Diamond-coated hollow drill.

When drilling deep holes, the drill bit should be retracted often and the drilling area cooled with compressed air. For *easily smearing* materials such as PE and PP, a very high feed rate is useful at slow drilling speeds. This allows the *frictional heat* to be removed with the chip.

For holes ranging from 10 to 150 mm in diameter, diamond-tipped hollow drills (Fig. 7.4.4) are recommended for thin-walled workpieces of thermosetting resins. After drilling, the cut surfaces should be *sealed* with resin.

7.4.5 Turning

The *lathes* used for turning should be high-speed, *variable* over a wide range with regard to *spindle speed*, and equipped with *air and liquid cooling*. Depending on the material, cutting speeds can be as high as 500 m/min.

The *turning tools* can be of high-speed steel if only slight *wear* is to be expected. The *tool angles* are shown in Fig. 7.4.5. For *thermosets* and, above all, *glass-fiber-filled plastics*, the tools should be carbide-tipped.

If the *cutting edge profile* is rounded slightly, no sharp-edged grooves will form on the *workpiece surface*. The surface will be extremely smooth. More detailed information on the *cutting edge geometry* of the tools for longitudinal turning and surfacing is contained in Table 7.4.

For extremely high-quality surfaces, the cutting tool should have a *broad cutting edge* as shown in Fig. 7.4.6. For reaming, *special bits* as shown in Fig. 7.4.7 have proven useful, since the conventional *reamers* have problems with chip removal. *Radial* and *tangential form turning* is possible if the workpiece is sufficiently stiff. *Cutting to length* is possible with the conventional tools for all thermoplastic and thermosetting materials.

Table 7.4 Values from VDI Guideline 2003 for the Turning of Plastics

	Plastic		Abbreviation	Tool	α	γ	φ	v	s	a
Thermosets	Laminates and compression molding compounds	Organic fillers	Hp Hgw GFK	SS HM	5 to 10	15 to 25 10 to 15	45 to 60 45 to 60	to 80 to 400	*	**
		Inorganic fillers		HM Diamond	5 to 11	0 to 12	45 to 60	to 40		
Thermoplastics	Polymethyl methacrylate and copolymers		PMMA AMMA	SS	5 to 10	0 to (−4)	app. 15	200 to 300	0.1 to 0.2	to 6
	Polystyrene and styrene-acrylonitrile copolymers Acrylonitrile/butadiene/styrene copolymers Styrene butadiene copolymers		PS SAN ABS SB	SS	5 to 10	0 to 2	app. 15	50 to 60	0.1 to 0.2	to 2
	Polyoxymethylene Polyacetal		POM	SS	5 to 10	0 to 5	45 to 60	200 to 500	0.1 to 0.5	to 6
	Polycarbonate		PC	SS	5 to 10	0 to 5	45 to 60	200 to 300	0.1 to 0.5	to 6
	Polyfluoroethylenes		PTFE	SS	10 to 15	15 to 20	9 to 11	100 to 300	0.05 to 0.25	
	Polyvinyl chloride (rigid) and VC copolymers Cellulose ester		PVC CA CAB	SS	5 to 10	0 to 5	45 to 60	200 to 500	0.1 to 0.2	to 6
	Polyolefins Polyamides (nylons)		PE, PP PA	SS	5 to 15	0 to 10	45 to 60	200 to 500	0.1 to 0.5	to 6

v = cutting speed in m/min
s = feed rate in mm/revolution
a = depth of cut in mm

* feed rate s = 0.05 to 0.5 mm/revolution depending on chucking and stability of the workpiece as well as the machine and surface quality

** depth of cut a up to 10 mm depending on chucking and stability of workpiece and the machine

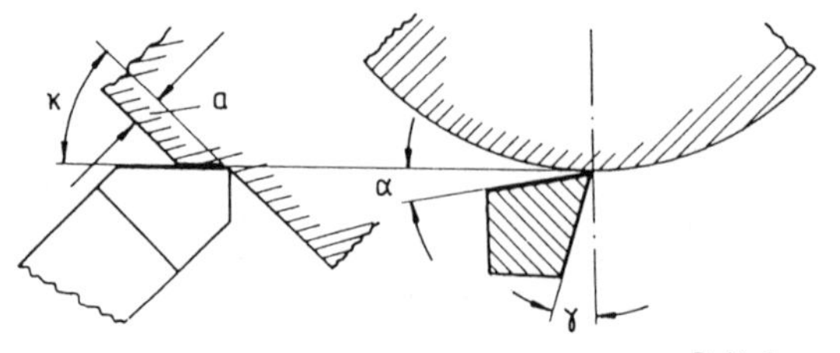

Figure 7.4.5 Tool angles for turning tools.

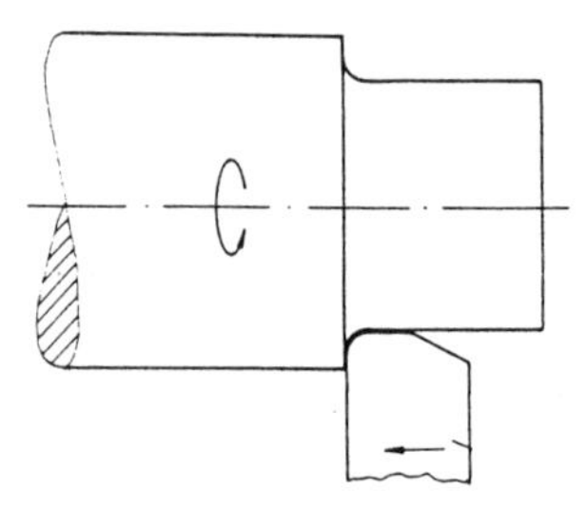

Figure 7.4.6 Tool with broad cutting edge.

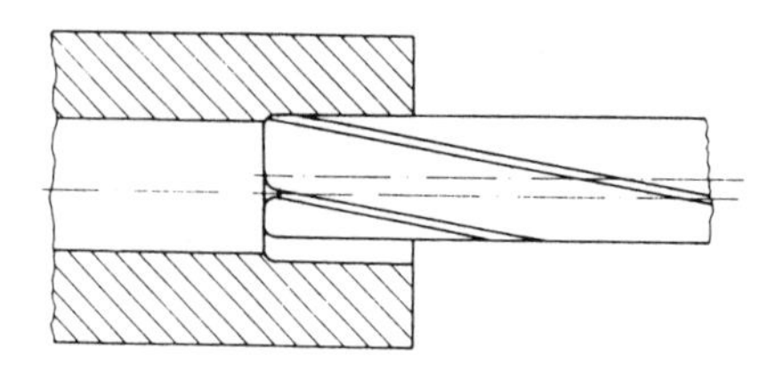

Figure 7.4.7 Special bit (reamer).

Bibliography for Chapter 7.1

Neitzert, W.A.: Preßluftformung von thermoplastischen Kunststoff-Folien, Rudolf Zechner Verlag, Speyer, Wien, Zürich, 1962

Thiel, A.: Grundzüge der Vakuumformung, Zechner & Hüthig Verlag, Speyer/Rhein, 1967

Neitzert, W.A.: Vakuum-Formung von thermoplastischen Kunststoff-Folien, Zechner & Hüthig Verlag, Speyer/Rhein, 1967

Höger, A.: Warmformen von Kunststoffen, Kunststoffverarbeitung, Folge 18, Carl Hanser Verlag, München, Wien, 1971

Throne, J.L.: Thermoforming, Carl Hanser Verlag, München, Wien, New York, 1986

Bruins, P.F.: Basic Principles of Thermoforming Gordon and Breack Science Publishers, New York, London, Paris, 1973

Bibliography for Chaper 7.2

Menges, G., Thim, J., Kaufmann, H.: Lernprogramm: Technologie der Kunststoffe, Carl Hanser Verlag, München, Wien, 1981
Menges, G.: Einführung in die Kunststoffverarbeitung, 2nd ed., Carl Hanser Verlag, München, Wien, 1979
Schwarz, O., Ebeling, F.W., Lüpke, G., Schelter, W.: Kunststoffverarbeitung, Vogel Verlag, Würzburg, 1981
N. N.: Kunststoffe, Schweißen und Kleben, Taschenbuch DVS-Merkblätter und -Richtlinien, volume 68/IV, DVS-Verlag GmbH, Düsseldorf, 1990
Potente, H.: Fügen und Umformen von Kunststoffen, Lecture notes RWTH Aachen, 1981

Bibliography for Chaper 7.3

Menges, G., Thim, J., Kaufmann, H.: Lernprogramm: Technologie der Kunststoffe, Carl Hanser Verlag, München, Wien, 1981
Fauner, G., Endlich, W.: Angewandte Klebtechnik, Carl Hanser Verlag , München, Wien, 1979
Menges, G.: Einführung in die Kunststoffverarbeitung, 2nd ed., Carl Hanser Verlag, München, Wien 1979
N. N. Klebstoffe und Klebverfahren für die Kunststoffe, Hrsg. VDI-Gesellschaft Kunststofftechnik; VDI Verlag GmbH, Düsseldorf, 1979
Schwarz, O., Ebeling, F.W., Lüpke, G., Schelter, W.: Kunststoffverarbeitung, Vogel Verlag, Würzburg, 1981
N. N.: Kunststoffe, Schweißen und Kleben, Taschenbuch DVS-Merkblätter und -Richtlinien, volume 68/IV, DVS-Verlag GmbH, Düsseldorf, 1990
Potente, H.: Fügen und Umformen von Kunststoffen, Lecture notes RWTH Aachen, 1981
Habenicht, G.: Kleben Grundlagen, Technologie, Anwendungen, Springer Verlag, Berlin, Heidelberg, New York, Tokyo, 1986

Bibliography for Chaper 7.4

Menges, G.: Einführung in die Kunststoffverarbeitung, 2nd ed., Carl Hanser Verlag, München, Wien, 1979
Reichenherzer, R.: Spanende Bearbeitung von Halbzeugen aus Kunststoffen, Werkstatt und Betrieb 104 (1971)
Schulz, H., Wiendl, J.: Eigenschaften, Herstellung und Bearbeitung faserverstärkter Kunststoffe, Werkstatt und Betrieb 117 (1984), No. 4
Schwarz, O., Ebeling, F.W., Lüpke, G., Schelter, W.: Kunststoffverarbeitung, Vogel Verlag, Würzburg, 1981
Saechtling, H.: Kunststoff-Taschenbuch, 25th ed., Carl Hanser Verlag, München, Wien, 1992
Spur, G., Wunsch, U.E.: Drehen von glasfaserverstärkten Kunststoffen und Schichtpreßwerkstoff mit PKD-Werkzeugen, Kunststoffberater No. 32, 1987 *VDI 2003*
Wagner, Fern: Kunststoffe in der Praxis, Verlag, W. Girardet, Essen, 1976
Zickel, H.: Das spanabhebende Bearbeiten der Kunststoffe, Carl Hanser Verlag, München, Wien, 1963
Zug, G.: Zerspanen von Kunststoffen in: Handbuch der Fertigungstechnik, volume 3/2, Spanen, Carl Hanser Verlag, München, Wien, 1980
Zug, G.: Beitrag zur Untersuchung des Zerspanverhaltens von thermoplastischen Kunststoffen beim Drehvorgang, Doctor theses at TU Berlin, 1970

Index

bstca 668 .4 M621